T0094149

Data Analytics Applied to the Mining Industry

Data Analytics Applied to the Mining Industry

Ali Soofastaei

CRC Press
Taylor & Francis Group
Boca Raton London New York

CRC Press is an imprint of the
Taylor & Francis Group, an **informa** business

1st edition published in 2021
by CRC Press
6000 Broken Sound Parkway NW, Suite 300, Boca Raton, FL 33487-2742

and by CRC Press
2 Park Square, Milton Park, Abingdon, Oxon, OX14 4RN

© 2021 Taylor & Francis Group, LLC

CRC Press is an imprint of Taylor & Francis Group, LLC

ISBN: 978-1-138-36000-6 (hbk)
ISBN: 978-0-367-61224-5 (pbk)
ISBN: 978-0-429-43336-8 (ebk)

Typeset in Palatino
by codeMantra

Contents

Preface

One of the initial landmarks in human civilization is mineral resources production. From the Stone Age to the Bronze Age and the Iron Age, our ability to innovate in our extraction processes for the most useful elements has been developed. It is becoming increasingly difficult to access and extract minerals. Mining costs increase due to rising labor costs and energy prices. To have a sustainable and affordable industry, there is no other way but to use new technologies. In order to reduce time and energy consumption and manual efforts to finalize mining projects effectively, computers and machines have been developed. Although the use of computer software in increasing the quality and reducing production costs in mines is widespread today, achieving better results requires the use of Artificial Intelligence (AI) and Machine Learning (ML) in this industry. Although today, all managers in the mining industry believe that they should use AI and ML, most of them do not know the correct way to use this science. Other than that mere AI scientists at universities do not have full access to the industry problems as well as the related data. This book potentially can make a bridge between developed knowledge by scientists at universities and research centers and industrial researchers at mining companies. The ultimate objective of making the relationship between scientific knowledge and industrial experience is to make the machines learn intelligently to think and evaluate the same thing as humans in different situations in mining operations. In the past, computers were doing as expected, but the systems now can think and behave like human beings with AI. High-tech giants are highly involved in research to develop the knowledge which has started bringing an innovative transformation. Although it is going to form the future mining industry, we need to know how it is affecting the mining workstyle. This book has been completed in order to give a glimpse of the application and advanced analytics and AI and ML in the mining industry.

Structure of the Book

Ten chapters have been designed for this book aimed to transfer the main part of practical advanced analytics knowledge to the researchers who are studying and working in the mining industry. All the presented information is supported by practical examples and scientific details. The chapters contain enough information for beginners to get familiar with the high technology and science application to solve mining business problems and more detailed technical information for advanced readers.

In Chapter 1, an initial review briefly gives a background of the digital transformation of mining. Modern technology is growing very fast, and businesses must adjust to new changes. In helping companies to this industrial revolution, digital transformation plays an essential role. The development of digital technology such as automation, sensors, advanced analytics, smart systems, etc., has compelled companies to consider new technologies that are more productive and efficient. As one of the leading industries in many countries, the mining industry faces a significant challenge when the sophistication of human, technical, and management systems is inadequate to open the doors to the old industry of the modern world. Digital transformation is a significant challenge. In order to address this challenge, the mining companies should work hard to meet this goal, as well as the universities, to develop the learning and research programs for potential mining engineers in the mining department. This chapter considers the mining industry's need for digital transformation and presents a three-part review of the principal elements of digital transformation, including data, connectivity, and decision-making. At the end of this chapter, there is a summary of both the mining industry and academic research perspectives on digital transformation to benefit mining firms.

Chapter 2 is about using advanced data analytics in the mining industry. The mining industry faces massive amounts of data that have hidden layers of information and knowledge. In addition, it is difficult for the industries to effectively and efficiently implement the data generated by their format, size, variety, and speed. Complexity in data processing and interpretation allows enterprises to use advanced technologies to solve raw data management problems. Big data analytics is a ground-breaking approach to data management. It uses machine learning (ML) and artificial intelligence (AI) methods to take advantage of the data that are collected. Chapter 2 consists of technical discussions regarding some mostly used ML and AI techniques in the mining industry. The presented discussions in Chapter 2 cover big data analytics, deep learning, and also machine learning application in the internet of things (IoT).

The realistic data collection, storage, and recovery technique in mining companies will be discussed in Chapter 3. To explore all the big data's potential and relevant technologies, basic data principles need to be thoroughly understood. The multiple available data sources and interdependencies between them need to be understood before any process of data analysis begins. In the composition of the business context and, therefore, the aims of the analysis program, different types, formats, and magnitudes of data are essential factors. This chapter opens a door in front of mining researchers to think deeply about the type of data, source of data, critical performance parameters, data quality (assessment, strategies, and improvement), data acquisition, data storage, and data retrieval in the mining industry. This chapter covers geological, operational, geotechnical, and mineral processing data.

The objective of Chapter 4 (making sense of data) is to create a data preparation framework to be used as a guide and best practices supporter for the adoption of data mining in the mining industry. Part I reviews essential aspects of data collection transition to data preparation and provides a summary of sources of data in the mining industry; Part II outlines steps, techniques, and issues to prepare data before analysis and modeling; finally, Part III provides extended data preparation considerations and applications for specific cases. Data analysis might seem a very technical activity at first glance, but with little guidance, every analyst and decision-maker can become a "data literate" and start mining datasets. Precisely, this chapter intends to play this role of guidance. However, it is essential to say that the chapter does not explore all the possibilities of data preparation; instead, the main goal is to generate initial interest in exploring data; for those readers who are interested in excavating their information and knowledge, a universe of material, articles, and references are yet to be explored in this (data) mining journey.

Chapter 5 presents the most used analytics tools in the mining industry. Presently, advanced analytics is a critical component of successful businesses in various industries. Mining plays a leading role in the development of the other industries and is rapidly developing in this industry with the help of the analytical tool. Many kinds of analytics have been discussed in theory. Nonetheless, choosing a practical instrument requires industrial experience and adequate competence in the knowledge involved. This chapter attempts to explain some practical analytical tools that address the problems of the mining industry. An introduction discusses the concept of each method, and the appropriate usage is discussed separately. The toolkits included in this chapter cover statistical and predictive approaches. The investigated predictive models in this chapter include the regression, time series, and machine learning methods. This chapter attempts to provide clear insights into the selection of the best analytical instruments for researchers to have better thoughts.

Process analytics is an essential practice for companies in order to deliver high-standard services or products to their customers. These technical analytics play the primary role in the mining industry, and the quality of data analytics is directly related to the accuracy of mining processes analytics. Chapter 6 explains more details about mining operational analytics and the importance of analytics to improve prediction, optimization, and making decisions. Traditional analytics approaches that are fundamentally developed around process data, such as Lean Six Sigma and business process analytics, are facing several limitations when confronted with the challenges of the big data era, characterized by real-time, high speed, dynamic changing, and multivariate requirements. Those methodologies can reach the next level by incorporating modern big data analytics techniques and technologies to boost their analytical power. Both literature and industry are full of real-case applications that support the introduction of big data analytics as game-changing technology in the process analytics – and improvement cycle. Studying this chapter is recommended to the mining researchers who are interested in applying advanced data analytics method to solve the practical mining business problems.

The authors technically discuss the predictive maintenance of mining machines by using advanced data analytics in Chapter 7. The way mining machines operate is dependent on the production of mines. It is, therefore, essential to maintain them. For an extensive mining transport system, the maintenance process is extremely demanding because it consists of many components. Maintenance techniques in mine sites exist in different forms. Prevention, failure, and predictive groups may identify mining maintenance strategies. Since the machines' reliability depends on several variables, it is not possible to fix the repair time for each component beforehand. Therefore, predictive maintenance is the most appropriate method. This approach provides continuous information on the state of the analyzed unit, thus monitoring the deterioration process and allowing the most appropriate duration of repairs to be scheduled. In the mining industry, development in online and standard acquisition systems is currently popular. The predictive maintenance today relies on the use of data fusion to continuously analyze data obtained from various machines in real time. It is necessary to suggest a set of time series indicators for the management and maintenance purposes that allow for a full and objective evaluation of the artifacts in terms of technology, economics, and organization, as well as an estimation of the remaining life of the artifacts. This type of analysis is a big data solution on an industrial scale. Consequently, the appropriate techniques for data analysis must be applied.

Fuel consumption and greenhouse gas emissions are two primary critical challenges in front of mining companies. The application of advanced data analytics to increase energy efficiency, reduce fossil fuel consumption, and consequently decrease the gas emissions in mines are the main subjects that will be discussed in Chapter 8. Different activities like drilling, manufacturing, transport, research and processing, and coal mining use much energy and release greenhouse gases. A better control of processes can substantially reduce the fuel consumption and gas emissions. The mining technique and the equipment used to determine the kind of source of energy in any mining activities are discussed. Mines and machinery for deciding the type of power source in any mining operation are studied. These machines, according to the production capacity and site layout, are haul truck excavators, diggers, and loaders, and use considerable amounts of fuel to operate in surface mining; therefore, mining is encouraged to conduct specific research projects on the energy efficiency of mobile equipment. Classical approaches widely used for energy efficiency and gas emission reduction are inadequate. The application of deep learning models and artificial intelligence is expanding in different industries and is a new revolution in the mining industry. This chapter gives an overview of the use of artificial intelligence technologies to predict and reduce the use of energy and greenhouse gas emissions in mines.

Decision-making is the last level of maturity in data analysis, and it will be discussed in detail in Chapter 9. This phase completes the analytics process that starts by gathering the data. Improving the quality of decisions affects the efficiency in the mining industry. Making the decision is one of the critical skills that mine managers need to lead the operation and maintenance teams effectively. Data analytics can also potentially help the managers to make better decisions related to safety, energy efficiency, final product cost, etc. This chapter explains in detail the effect of advanced analytics to improve the quality of managers' decisions in different situations. After a short introduction, the organization design and KPIs are explained, and then, the advanced analytics role in making practical solutions is clarified. At the end of this chapter, the expert systems components, types, and methodologies, especially in mining, are explained.

Chapter 10 presents a useful discussion regarding the future skills that the mining industry needs. The science and technology are growing very fast, and industry leaders must make comprehensive plans to transfer new knowledge and updated technology into their companies. This approach not only can help the industry to increase the efficiency in different areas but also potentially can help societies to have a sustainable environment. The application of advanced analytics and using innovative methods such as machine learning, artificial intelligence, and deep learning algorithms are pioneer technologies to the current industrial revolution. If we are interested in having a successful digital transformation plan in the mining industry, it is essential to make sure we not only can update the employees but also we have an acceptable training system to prepare the new workers, managers, and decision-making people for the future. This dream will be made real when there is a functional relationship between universities and companies. The mining industry, as a critical industry, plays a critical role in training the new miners for the future mining projects globally. The chapter presents some new requirements, skills, and related training programs in the mining industry.

This book tries to help readers to have a better vision of advanced analytics in the mining industry, and the authors hope that this volume will be a valuable resource for industry professionals and researchers. The presented chapters in this volume signify state-of-the-art regarding critical topics in advanced analytics, machine learning, and artificial intelligence. The breadth of coverage and the depth in each of the sections make it a useful resource for all mine managers and engineers interested in the new generation of a data analytics application. Above all, the author hopes that this volume will spur on further discussions on all aspects of advanced data analytics applications in the mining industry.

Dr. Ali Soofastaei
Global Projects Leader, Artificial Intelligence Center, Vale
Australia

MATLAB® is a registered trademark of The MathWorks, Inc. For product information, please contact:

The MathWorks, Inc.
3 Apple Hill Drive
Natick, MA 01760-2098 USA
Tel: 508-647-7000
Fax: 508-647-7001
E-mail: info@mathworks.com
Web: www.mathworks.com

About the Author

Dr. Ali Soofastaei leads innovative industrial projects in the field of artificial intelligence (AI) applications to improve safety, productivity, and energy efficiency, and to reduce maintenance costs. He holds a Bachelor of Engineering degree in Mechanical Engineering and has an in-depth understanding of energy management (EM) and equipment maintenance solutions (EMS). The extensive research he has conducted on AI and value engineering (VE) methods while completing his Master of Engineering also provided him with the expertise in the application of advanced analytics in EM and EMS. Dr. Soofastaei completed his Ph.D. at The University of Queensland (UQ) in the field of AI applications in mining engineering where he led a revolution in the use of deep learning (DL) and AI methods to increase energy efficiency, reduce operation and maintenance costs, and reduce greenhouse gas emissions in surface mines. As a Postdoctoral Research Fellow, he has provided practical guidance to undergraduate and postgraduate students in mechanical and mining engineering and information technology. In the past fifteen years, he has conducted a variety of research studies in academic and industrial environments. He has acquired in-depth knowledge of energy efficiency opportunities (EEO), VE, and advanced analytics. He is an expert in the use of DL and AI methods in data analysis to develop predictive, optimization, and decision-making models of complex systems. Dr. Soofastaei has been involved in industrial research and development projects in several industries including oil and gas (Royal Dutch Shell), steel (Danieli), and mining (BHP, Rio Tinto, Anglo American, and Vale). His extensive practical experience in the industry has equipped him to work with complex industrial problems in highly technical and multidisciplinary teams. Dr. Soofastaei has more than ten years of academic experience as an Assistant Professor and leader of global research activities. Results from his research and development projects have been published in international journals and keynote presentations. He has presented his practical achievements at conferences in the United States, Europe, Asia, and Australia.

1

Digital Transformation of Mining

Ali Soofastaei

Introduction

Adapting the mining industry with the technological changes is an exciting research subject [1,2]. Studied research about sociotechnical theory in an Australian mine site shows one of the first experiences to transition from hand-got mining to longwall methods. As a practical experience, this study illustrates a successful transition from a traditional mining method to an advanced process when socio-psychological and production influences grew over this technology transition [1].

The fourth industrial revolution was happening when the world is facing with the digital decade [3]. In mining, a massive amount of data is collected from many equipment and machines working in the sites that are much more than ever before [4]. These data can potentially make excellent opportunities for mining innovation to find new solutions for business problems through digital transformation (DT) in this industry [5]. The main goal of DT programs in mining has come to describe how companies become accustomed to digital modifications [6–13]. Moreover, there is not the same definition of digital mining transformation [13]. Figure 1.1 demonstrates a technology-driven process consists of three main components of DT: data, connectivity, and decision-making [9].

A successful DT plan can increase the digital capabilities and develop the sociotechnical capacity in a mining company [2,14]. DT can also change all aspects of the business to improve the mining operation and maintenance [15]. However, mining companies are struggling to start the DT plans based on their technical and management challenges they are practically facing.

The pressure on mining companies to adopt themselves with digital technologies is on both sides: supply and demand. In general, the trouble starts on the side of the consumers. Some examples of this pressure are explained as follows:

Influential factors on demand:

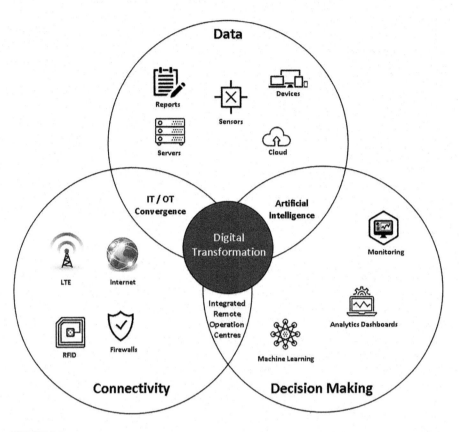

FIGURE 1.1
DT components.

- Consumers are more connected and are more significant decision-makers.

  The digital economy produced a cultural transformation that has set a higher level of expectation and user experiences from consumers. This change redirected the decision-making from the mining companies to the final consumers.

- Consumers are more focused on user experience than with the possession of the property itself.

  New business models developed by the digital economy lead a transformation in the consumers' preferences mainly among the young generation, migrating the focus more and more from owning to using.

- Liquid expectations.

  The more developed a digital economy is, the more consumers extrapolate the consuming experience of a determined category of mining product to other markets, thus significantly amplifying what the market traditionally defines as "competitor." Currently, competitors are not necessarily inside the mining industry.

- Faster adoption cycles of new ideas and technologies have made markets quickly disappear.

    The classic curve of mining innovation diffusion is facing a significant change. The process of transmission that once slowly flowed between the social systems participants nowadays quickly converges between the winner solutions.

Influential factors on supply:

- They are unbundling phenomena by the start-ups.

    The entire process of a productive chain, which was executed before for a big mining company, currently can be achieved by hundreds of small companies that perform each one of the small steps of the whole process in a more efficient way.

- Exponential cost reduction of the technological process.

    This pattern, which has been observed since the end of the fifty decades, has become economically feasible in a series of projects that previously did not leave the drawing board.

- New competitors being created every day.

    It is essential to plan a DT plan to predict the effect of market conditions on the mine value chain. The companies that do not review their operational models and especially their business models will not have space in this dynamic competitive environment. This DT plan can be reached through three strategical drivers:

    - Digital Business Transformation

        Attending the new demands of business models. The primary investment area to implement this strategical approach is a junction of the technological parks with the relevant set of new and existing data to foster the use of machine learning and artificial intelligence (AI). This approach can help to identify new trends and market demands.

    - Digital Clients Transformation

        Revision of the client experience B2C or B2B. The integration of different platforms to guarantee clients information unification, jointly with the DT of the marketing function, is the necessary condition to implement this strategical driver. The application of AI unitedly with mobile technologies and social media is essential to customizing the offerings to guarantee higher client engagement.

    - Digital Company Transformation

        Operational excellence of production process and technological park. Each productive process automation is required to implement this part of the strategy, which ranges from the operation itself to the system for decision-making. The use of IoT, robotics, and AI are some of the elements that allow automation and identification of opportunities for an improved efficiency.

To maximize the return of investment in digital is required to focus on some leveraged strategies; on the contrary, the train of DT takes the risk of stopping at proof of concepts and the first results never turning out to be sustainable.

- **Agile leadership**
  Strategic view and fast-paced in the decision-making process.
- **Workforce focused on innovation**
  Digital mindset infused in the workforce
- **Network**
  Keeping the mindset of ecosystem collaborating inside the value chain (e.g., suppliers, logistics, clients) and outside (e.g., start-ups, universities).
- **Access, management, and usage of data**
  The capacity for creating knowledge to improve the decision-making process.
- **Appropriate technological infrastructure**
  Guaranteeing processing capacities, data and business security, and interoperability among systems.

In the past, the mining companies could choose to be later/early adopters regarding the new technologies. However, this is no longer the reality nowadays, considering what is highlighted above, and the journey on DT becomes an essential plan for all companies working in the mining industry.

Overall, there are four highlighted summaries for DT in the mining industry as follows:

1. Mining companies should start DT program as an essential revolution in this industry.
2. There are three foundational components of the digital mining transformation process. These components are information, intranet and internet connectivity, and decision support.
3. DT delivers a conversation on how it will be an essential part of the achievements of mining businesses into the future era.
4. DT recognizes the strategic fields in which organizations of higher learning can supply the required resources to support the mining industry.

There are some suggestions to have a successful progress in the DT journey in the mining industry. First, it is essential to define a clear responsibility for digital investments. Second, companies should invest in use cases, not just in technologies. Third, it is necessary to use the result-based actions according to the theoretically designed approach. Fourth, the companies take full advantage of the low-hanging fruits, i.e., low cost and fast, successful opportunities, which will help the company to create a digital culture. Fifth, take risks in assessments that identify common problems of several company sectors. Therefore, it will be easier to scale and reuse the lessons learned. Sixth and on a final note, the suggestion to have a successful DT plan in mining is thinking about a multidisciplinary concept, and this approach needs innovative discoveries in all company sectors.

DT in the Mining Industry

A short industry review shows the need for DT and challenges. The mining industry needs to critically use DT to increase the safety, productivity, and efficiency. However, this industry is behind most other industries, as shown in Figure 1.2.

A completed review of the yearly published documents from the industries' top ten mining companies shows that six out of ten stated that DT is a part of its policy, three out of ten corporations list qualitative consequences from digitalization programs, and only one out of ten might provide quantitative value for the benefits of DT [17]. However, this story has been changed, and currently, there are many other completed successful DT plans in big mining companies globally.

Just as in cases where mining businesses are running to have achievements from DT programs, there stays a substantial need for employees with the essential unconventional skills to execute them [18]. Technically, it is difficult to assess what precisely every mining business is performing

Relative Digitization

Low			High
Mining	Health	Oil and gas	Media
Entertainment	Education	Real estate	Information technology
Construction	Chemicals	Transportation	Finance and Insurance
Agriculture	Good Manufacturing	Retail trade	Utilities

FIGURE 1.2
Relative digitalization by industries [16].

for DT. However, the overall developments for the industry may be created by exploring the regularity of related terms as they publish in annual businesses' statements. Yearly reports for forty-one of the biggest global mining companies were investigated for some general conditions concerning information, analytics, people, and technology (see Figure 1.3).

Figure 1.3 demonstrates the number of cited terms related to the DT in some published annual reports by top mining global companies. The terms technology and data exist for the most part in the reviewed documents.

Nevertheless, these phrases were frequently cited out of the perspective of DT. Creativity and Innovation were cited very regularly and ordinarily close related to the term technology, which indicates that technology and innovation are the foremost popular concern through top mining businesses. In this investigation, the term cyber has been mentioned 132 times, and generally from the perspective of cybersecurity, it is becoming important for approximately half of the businesses. Autonomous and automation and merged were cited about as much as cyber. More than half of the mining companies mentioned information technology. However, operations technology has been stated just once by one business. After All, DT has been cited by three businesses for a full amount of eleven times. The small number of companies quoted from DT may mean that companies, which do not know what DT is, choose to use different stages in order to convey a similar concept or do not regard DT as something worthy of its shareholders.

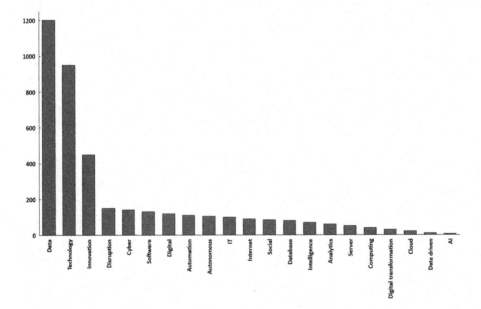

FIGURE 1.3
Regularity of DT-associated terms in mining companies' annually published reports [18–24].

A completed assessment across all business sections in mining operations shows that over 80% of them have essential DT initiatives in their minimum and long-term plans [25]. Besides, 40% more companies with practical digitization approaches are expected to find new, unexpected ways to improve the effectiveness of their businesses. The new digital funds for mining are eight times larger than in the industry five years ago, and projects are completed in 50% of the time, for these companies in contrast to their peers [11]. In several other mining fields, DT derives profits from the production of consumer-focused goods and that increases customer understanding [13,26]. However, other mining companies have many customers, and the products of mining provide many additional benefits before reaching the end user. This is a financial opportunity to improve customer engagement for mining companies. For the sake of a broader section of the mining community, diversity, geotechnology, and digital awareness are directly related to the mining company's ability to implement the DT programs [27–30]. Around 32% of the working population in mining were suppliers [31], digital governance (DGs) only confuse as mining companies have less influence than their staff over providers. This means a few years' retirements, workforce reduction, lack of skills, and also has a significant impact on the ability of the industry to respond to the changing technological transformations [32].

Diverse mining companies rely on the techniques of enterprise resource planning (ERP) as an alternative to the implementation of DT strategies [33–35]. ERPs are essential to set procedural standards, but they do not establish organizational comprehension or competitive advantages [35–38].

In addition, ERPs allow multiple companies to acquire DT solutions off-the-shelf. These systems and software products, however, will necessarily not adhere to the DT design, including introducing innovation models and a new culture of management of information management [6,8,13,18,25,39].

Data Sources

The difference between Information Technology (IT) and Operations Technology (OT) needs to be clarified in the mining industry. Data are considered an asset, and today, having access to the massive data quickly makes a new challenge named big data [40–47]. Big data are a challenge, but it can be a great opportunity to develop new know-how to deal with business problems. High-quality electronic circuits with integrated self-observation and intelligent methods are economically low and result in an extensive exchange of information in mining [48]. Big data are shown that suit variety, speed, volume, and accuracy [45]. Variety means multiple ways of producing data. Speed means that the information is evolving in a dynamic environment.

A large volume implies a large amount of data. Veracity reveals significant numerical and noise errors. Most knowledge on my web is obtained from OT and often set as logged data. The relational database management (RDBM) periodically handled structured information and the application query language (SQL) and similar database management systems [49–52]. In addition, organized information can be time-series data, often from signal-based data, and managed by data historians [53,54]. Similarly, standardized information for one-time encounters in preliminary information is also different; for example, the specifics of the fleet management system (FMS) [55,56] may want to have outliers eliminated during cycle times as a result of FMS errors. Unstructured information is available in various ways, and without alternatives or multilateral channels, it is very difficult to operate [57]. Useful information in any system depends on robust data structures to guarantee its speed, accuracy, and testability [40].

IT is generally all about computer technology, including networking, hardware, software, and the Internet. OT usually refers to software and hardware that regulate and control physical objects, procedures, and measures within a business [18,58]. In addition, the OT methods used by an institute in its workflow movements are characterized [59]. There are strong ties between an association's implementation of new information and subsequent structural adjustment [8]. Also, OT frequently includes the department or team within the company responsible for OT study, maintenance, development, performance, and management [60].

DTs have merged the OT core with the enhancement of software and introduced operational prototypes both for the operator and for the business. Strictly regulated and significant worldwide operations pose many risks [61], and risk minimization includes the acquisition of skills between OT and IT expertise [8,62].

The IT framework is not intended to include proven expertise that creates information primarily for mining companies [18]. OT and IT typically played individual roles in mining companies. A wide range of emerging technologies such as engine to the machine, networked sensors, neural networks, cloud machines, machine learning (ML), predictive analysis, optimization technologies, policy maps, edge computing, wireless networking, and the development of Internet-capable applications, including monitoring systems, has been integrated into the connected architecture of OT and IT [62,63].

Figure 1.4 displays the results of finished inquiries into the performance of OT and IT cooperation [62]. The research included 151 people who worked for energy companies using the industrial management techniques.

The results shown in Figure 1.4 indicate that the OT-IT relationship is both in quantity and in improving efficiency. The survey found that 73% of the respondents now accepted a more reliable and more effective relation; 19% of the IT sectors surveyed believe it is currently weaker than in the previous survey [62]. The survey also found that it was challenging to discover new talent in both the OT and IT industries.

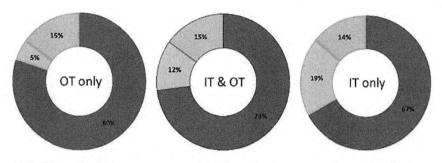

Collaboration became stronger **Collaboration became weaker** **No change**

FIGURE 1.4
IT and OT teams work [62].

DT also has increased access to human-generated information in mining, such as preshift risk assessment records, ear charts, explosive management reports, pit plans, block models, and maintenance task orders. Workers may also receive information at mines to verify, for example, whether their eyes are opened by the facial features of mobile equipment drivers [64]. Already, heat stress data for miners are available in the dry underground environments [65]. New technologies have many opportunities in this field to enhance information collection [65].

Currently, there are new data sources in the mining industry. Uncrewed aerial vehicles (UAVs), self-directed processes, smart equipment, robots, metadata (information generated as a consequence of information analysis and exchange), and wearables are including the new information resources which will need mining processes to adopt better DT approaches [66–68]. OT's advance in mining companies raises information quality and quantity. Having access to the latest datasets include photogrammetric information from UAV surveys, wearable in-house staff in real-time, on-belt sensor ore feed quality, or equipment performance information with an incomparable granularity [47,52,69]. Modern mining also has real-time mining environment monitoring sensors such as moisture, temperature, and gasses for underground or sloping mines, weather patterns, seismic, and boiling controls for surface mining, remote machine safety, repair, and substantive operation of equipment [70].

Connectivity

Connectivity is one of the essential components of the DT. Connectivity is broken down into six subjects, involving data knowledge exchange, the IoT, cybersecurity, integrated platforms, wireless communications, remote operation centers, and 5G technology.

Information of Things (IoT)

The enormous network of connected physical devices is generally mentioned as the IoT. Figure 1.5 displays the IoT mining landscape.

Commercial IoT reflects IoT (IIoT) in the commercial domain. However, the IoT and IIoT words are utilized exchangeable regularly [71]. IoT is generated by a managed information flow process from the built-in machines and OTs to data storage facilities and analytical platforms, where data flow to users [48]. As shown in Figure 1.5, mining companies can attach processing machinery, heavy equipment, staff, and sensors to connected platforms by using IoT. Such integrated frameworks contribute to better decision-making through data/information analyses.

Data Exchange

Collecting enormous information is not helpful [51]. A database can be developed, organized, grouped, and defined to support business analysis and decision-making processes, including information from multiple sources [7,37,50,72–75]. Data storage systems act as organizational repositories of information that can be commonly used by an entity to test and analyze a variety of parameters of interest, such as the performance of motor components shortly before failure. Data for the engine components could be generated from an unconnected and remote part of the company. However,

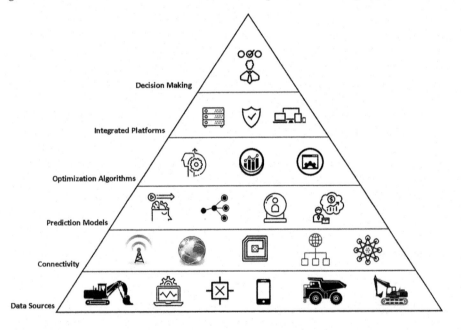

FIGURE 1.5
IoT in mining IoT landscape [38].

they could be transmitted into a database where a valuable analytical tool could be created, used anywhere within the company, possibly far beyond where the engine is connected [76]. Instead of lying on a file, information gathered can be of interest by supplying the consumer with real-time data about the demands of the saleable commodity before it reaches its destination. The extract, transform, and load (ETL) procedure represents up to 80% of the work required for the generation of a data store [50]. ETL includes the following subjects.

- detecting data from its primary resource,
- generating applications to obtain, fill, and adapt the information to a general structure, and
- inserting it into the data store.

Different companies use martial data as a reduced method for data storage, depending on the nature of data in data stores. Data marts are small data warehouses that provide decision support systems (DSS) for a limited number of operators [77].

Safety of the Cybers

Approximately 18 billion IoT machines are expected to be available in various industries in 2022 []. IoT devices are being populated and used so rapidly that the issue of how they affect cybersecurity is becoming more and more apparent [61]. In general, given the potential cybersecurity risks, these devices are supplemented because it can minimize mining companies' operational and maintenance costs. This is true, especially with small embedded devices, which are only newly built to be smart. For these machines, safety is frequently different from the traditional ones, which follow more predefined protocols of communication. Cloud services complicate the security problem of those industries that have previously been restricted to a company's network and have internet connections [78].

Remote Operations Centers (ROCs)

The remote operating centers (ROCs) promptly develop essential mining tools for synthesizing, managing resources, and relying on vast numbers of inputs. Conventional techniques such as simple reports and tablets are not enough for implementing and analyzing information.

ROCs use long-term technology, such as data visualization and cloud computing, by companies outside the mining industry to ensure a realistic and practical approach to fast and impactful critical decisions [79].

Also, ROCs reduce the risk and boost the work experience of workers. Staff will be removed from risk by using semiautonomous equipment, thus eliminating the need for cost-effective and less reliable security systems, such as personal protective equipment (PPE), engineering controls, etc. The NIOSH hierarchy of controls offers the most effective and acceptable approach to the eradication of risk by employees who are exposed to hazards [80]. In ROCs, employees generally operate in secure, climate-controlled rooms and can monitor other equipment at the same time, thus significantly improving productivity, protection, and energy efficiency.

Platforms Incorporated

By integrating various OT and IT technologies, mining companies are turning their operations into information systems (ISs). Experiences with mining projects show that scalability is essential in every integrated platform.

Programmable application interfaces (APIs) provide a suitable and scalable solution for incorporating a variety of functional groups and information sources with a unique system in mining companies [81]. An API provides a protocol to ask for resources from an application for a developer or an external system [77]. An alternative can be linked, and data can be shared using the APIs as a multiplatform program. The field of information systems management, which generally had nothing to do with computers [82], experiments on the implementation and acceptance of IS by operators are currently underway [83]. The complicated environment of mining requires building recognition, labor problems, plant operations, data quality, the complexity of the workplace, inclusive knowledge components, and ongoing technical training for IS mining specialists to be successfully integrated.

By using the Internet Protocol, IoT turns the physical world into a kind of IS. By linking goods to information, this revolution removes the gap between data and materials. What is left is a complete restructuring of machinery and process management. Figure 1.6 demonstrates such a scenario as an interconnected network.

Since many mining companies still have an integrated platform, the term indicates and implies that integrated systems, equipment, machinery, and workflows merge into one unit.

Wireless Communications

Generally, mining sites are in remote areas, and their environment is extremely harsh. Furthermore, the mining operations are continually changing [61,84]. There are significant obstacles to large mobile devices' wireless interconnections at mining sites. These mobile tools have been used extensively for decades, but only for many years are network hardware companies supporting their devices until they become obsolete [61]. Miners need cable systems that can meet the different and versatile needs of their company over time. Besides, the risks to cybersecurity increase when data transmission volume [61,85] and networking specifications increase [86].

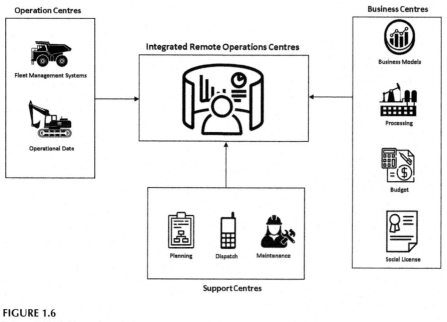

FIGURE 1.6
Integrated platform for mining.

Optimization Algorithms

Decision-Making

Decision-making is the last component of DT. Decision-making is a process of creativity involving both people and technology. Such decisions are defined as a support for data-driven decisions [74,75] or decision-making [7,72,73,75].

There is not any doubt that managing the data is critical for all businesses. However, there is considerable that turning the data into action is more important [41]. According to Harvard Business competition reports, OT is only appropriate for the business processes because it provides businesses with the "capability to act" [88]. Comprehension of data leads to practice. Historically, talent and experience have been a significant factor in mining decisions [73,87]. However, skill- or experience-based judgments are no longer satisfactory. It is critical for coordinating decisions based on knowledge and thinking directly and indirectly [88,89]. Consequently, the real benefit of DT is the determinations based on data. That is because small choices for an automatic system can be made automatically, but even these systems should optimize their extra benefit [90]. Much work has been carried out to ensure that machine technology is synergized with the intellectual abilities of people in operations, like human-computer interaction [83].

Advanced Analytics

Mathematical analytics can be categorized into three different groups. These classes provide analyses that can define, predict, or prescribe. Data are used by concise analysis to solve the problem [91]. Predictive modeling uses historical data to predict what will occur in the future. Prescriptive analysis generates methods for solving the problem of what should happen from both descriptive and predictive analyses. Segmentation, reportage, and predictive modeling are rising advanced research models [36]. Advanced analytics are most effective if the difference between business processes and raw data was removed. Many analytical attempts fail without this, and there is no point in trying to produce, store, and compile the data [41]. Advanced analysis preserves planned growth, offers a competitive advantage, and supports enterprises. Table 1.1 outlines how advanced analytics maintains planned development. Although advanced research affects the whole business, much like a whole project is used by a data warehouse, the functional analysis uses different silos in a commercial, such as data marts.

The practical advanced analytics companies' standard of the mining industry is presented in Table 1.2. These practical advanced analytics companies regularly assess various scenarios, involving investments, cash flow, reliability, probability, and maintenance [73].

Individuals

DT produces a collaboration of business analysts and data analysts in the process of the merger between IT and OT. The business analysts who have historically analyzed and detailed the rules and procedures of a company have traditional roles held by IT people to maintain the same rules and processes. The structured investigative response was designed to create a void for data scientists in organizations where none were previously developed with more constant data sources and more robust business procedures [92]. In addition to those changing roles, advanced work requires expertise, mainly based on digital learning. Digital learning means that information generated

TABLE 1.1

Advanced Analytics in Planning

Analytics Approach	Description
Wild & Hex's Delta Model	Establish a link with customers through the best invention, customer responses, and lock-in approach.
Porter's Five Forces	Planning for new market entrants, performance, financial rivalries, contractors, and customers.
SWOT	Planning about assets, difficulties, limitations, and opportunities.
Company view based on capital	Advancement VRIN supplies by valuable, unique, and nonsubstitutable

TABLE 1.2

Advanced Analytics in Mining

Category	Benefit
Human resource advanced analytics	Human capital management
Staffing and scheduling	Employee productivity
Supply chain analytics	Better management of supply networks
Commodity market analytics	Inventory and stockpile management
Geology and exploration	Exploration portfolio management
Mining and processing analytics	Reactivity, resilience, and execution of delivery
Safety and risk analytics	Risk management
Permitting and environmental analytics	Transparency
Financial analytics	Business management

from a range of websites can be recognized and used in several formats. Digital learning frees the public from the previous medium constraints and promotes the form of information generated for the public; it is incorporated [93]. Digital learning, unlike conventional learning, does not imply an awareness of what topic is ultimately created, but rather the possibility of creating and collaborating with a user-disposed subject. Digital learning is an expert group conventionally found among computer science graduates who are talented in the mining industry [94]. Besides, successful advanced research programs need to consider what the students need in metallurgy, mining, and geology. The best response is a team of multiple disciplines. Hackathons are the tool used by mining companies to recruit new, untraditional talent [95].

Process of Analysis

The advanced analysis approach is an iterative and a scalable analysis mechanism, as shown in Figure 1.7.

To investigating the industry problem understanding, the advanced analytics procedure starts by discovering a challenging area of the industry procedure that is not entirely known. The preparation of data will appear to overcome the challenge in handling each portion of the data science devices currently available [92]. All of these methods, categorization, regression, cause-forming, clusters, correlation, data reduction, cooccurrence, monitoring, link prediction, cross-relation, decision trees, and the identification of anomalies, were completed in some references [41,43–47,92,96].

Technology in Advanced Analytics

Some are open source, and some are proprietary technology. Table 1.3 shows some of the most common advanced financial data technologies.

Several of the accessible data-advanced analytics technologies utilize the Software as A Service (SAAS) business model. SAAS requires a digital

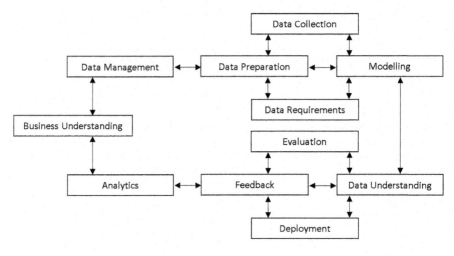

FIGURE 1.7
Phase of data management and analysis.

TABLE 1.3

The Most Famous Technology of Business and Open-Source Data Analysis

Application	Description
⁺⁺⁺ +ableau·	Collaborative information visualization and Business Intelligence (BI) application
Qlik Q	BI platform that discovers information understandings around various sources
MicroStrategy	Software resolutions to develop and implement analytics and flexibility applications
D3	Data-driven documents JavaScript library
§sas.	Application platform transmuting advanced information analytics into intelligence
𝒢𝜑 Gephi	An open-source information visualization platform
FLOW DATA	A society for data actors to recognize and share visualization methods
ARBOR NETWORKS	A Java-based graph library
geocommons	A society developing an open-mapping platform

(Continued)

TABLE 1.3 (*Continued*)

The Most Famous Technology of Business and Open-Source Data Analysis

Application	Description
	Analytics and Business Intelligence software
	Geographical information system
ac●uire	Geoscientific information management system
leapfrog	3D geological modeling application for the mining industries
dmsi	Provides multidisciplinary research for repair approaches
earthsoft	Advanced data analytics and decision-making support approach to compliance monitoring
esri	Geographical information system
VIZIYA	Working® Analytics is a robust BI system for data management for businesses.
OSIsoft.	Intuitive PI System Data Web-Client Visualization (PI System connects sensor-based information)
:RESOLVER	Risk management framework with risk management analysis and reporting ability
Dashboards	Visualization and data management software fully customizable
IMDEX ioGAS	An experimental data analysis application for metallurgical and geochemical data
PRONTO SOFTWARE	IBM Cognos-incorporated full ERP program
GEOSOFT.	The technology suite for the analysis and review of complex geoscience issues

(*Continued*)

TABLE 1.3 (*Continued*)

The Most Famous Technology of Business and Open-Source Data Analysis

Application	Description
BLUE MARBLE GEOGRAPHICS	Technology for GIS-based modeling and presentation
DOMO	Completely mobile, the cloud-based mobile BI solutions
Minitab	The Pennsylvania State University data and research kit
HEXAGON	Live dashboard and web-based logging and review
UPTAKE	OEM agnostic data analysis platform
COGNOS Better Decisions Every Day	Information visualization tools from IBM Research
JavaScript	Web devises to generate Interactive Data Visualizations

subscription to software products, which require requested responses, including the fact that companies themselves need the installation and maintenance of the computer software [44,95]. Therefore, most frameworks and computer software systems are reliable.

The ability to manage highly different data and computational load, scalability, additional software flexibility, parallel processing, and economic applications is included in data mining [92,94–99]. Dasgupta is a robust open-source platform for the storage and management of data generated by the Apache Software Foundation (ASF) [44,46]. A lot of programming languages, including R, Python, Spark, and MATLAB, are available [46]. These languages allow data analysts to use ML, and AI techniques that can add value to complex business problems through scalability [9].

DT and the Mining Potential

Because of its tempo, size, and complexity, DT is significantly different from previous IT transformations [6–13,66,97,98]. The potential benefits also differ significantly. Those benefits can be defined in three different categories: individuals, processes, and technology in the rest of this section.

The Role of People in Digital Mining Transformation for Future Mining

The latest studies show that roughly 1% of the global population are mining employees [99]. The DT, which includes the IS model cantered about human principles, will be an essential social obligation going ahead [83]. Learning systems have been tailored to digitally educated and very qualified analysts' requirements. According to the newly produced state of the university in the USA in 2019, about 90% of the high-ranking higher education institutions offer analytical programs. When the rise in analytical education persists, analysts will come into the mining business in more significant quantities.

As the IoT sector remains to expand, more analytical learning, IT, and OT are expected to be integrated into the university curriculum. Besides, the potential mining staff must sustain digital learning as a result of public learning [100]. Also, women are still entering mining, providing several advantages [27]. Such changes require a change in when, how, how, and the pace of decision-making. A further investigation of decision-making techniques and their relations with diversity will be necessary because the potential increase in the number of different monopolies is one of the most significant risks of the new digital approach [97].

Moreover, without access to digital improvements and innovations, people and countries can see even better employees and inequities in skills, developing "digital refugees" [101,102]. Mining companies that overestimate insights tend to be more open to digital changes and, therefore, profit [73]. Mining companies that overestimate intuitional options may not be as realistic and struggle to achieve the advantages of DT [88,89]. It can reveal leading indicators of diversity in an enterprise by examining decision-making approaches. Such vital metrics can be used to help measure and achieve diversity and inclusiveness. Also, as AI continues to be developed, companies that can leverage their potential will benefit, especially when dealing with difficult challenges [72,103,104]. AI technology will most often be accomplished, including principles, stakeholders, structured research, and an insight into the future of AI [72,75].

The Role of Process in Mining Digital Transformation for Future Mining

Given that DT will change the mining industry, the most ground-breaking hope in the future maybe that one day, mill and site managers will sit at a similar table, not including confusing maintenance and operational details. By improving the remoteness, security, and automation of maintenance and operations, DT will change the required mining procedures. Short-term planning and monitoring of short intervals are the goals of change to DT because the decision and planning systems need actual data to be assisted. A new study on the short-term planning identified critical concerns and potential short-term planning guidelines that are enabled by DT [105]. Following the strategies of Mine, Moving, and Mill (M2 M), which removes the bottlenecks to the whole operation, as fuzzy patterns become visible [52]. Mining firms will gain better knowledge and employment with local communities through DT, which influences different media platforms.

The implementation of an organizational framework such as agile application development will help mining companies that change their structure. Agile development is based on four key values, namely, interactions and individuals on procedures and tools, working through comprehensive customer cooperation through contract negotiation, documentation, and reaction to subsequent plan changes [106]. The effectiveness of agile development in applications depends on twelve core assumptions: work motivated by business value, hand-on business outgoings, direct involvement of stakeholders, real estate deadlines, responsive preparation, self-motivation management, just-in-time communications, waste management, constantly monitored efficiency, mirror revision control, immediate progress tracking, and continuous improvement [107].

For example, market competition, resource depletion, and social dynamics can be controlled in an agile framework for the exceedingly complex and evolving parameters of the mining industry. Agile methods of production not only make it easier for organizations to accept changes but also increase the efficiency of repetitive processes [107].

The Role of Technology in Mining Digital Transformation for Future Mining

Based on some recent research, nonrelational database management systems, including NoSQL, Hadoop, Amazon Redshift, MPP analytics, and Google Big Query, now account for over 70% of analytical data. Some platforms are scalable and efficient with information analytics demands; there is a risk that the growth of certain systems will continue. These knowledge pieces will also change and reduce the amount of time it takes to integrate information.

Moreover, the industry has recently made virtual reality (VR) popular. VR is a technology with a large mining potential and is especially capable of improving remote control and autonomous equipment operation [52]. VR mines aid employee performance management by designing incidents that would otherwise not happen to workers or in unavoidable circumstances. The capacity for accident investigations is immense [108].

In addition, blockchain is another technology that disrupts the mining industry potentially. Within mining companies, we have at least nine cases of blockchains, including, data integrity, provenance, cradling-to-grave blockchain properties, IoT-based workflow automation, supply chain optimization, tokenized mining, healthcare, and staff management [109].

Blockchain also impacts industry and innovation, which can have indirect and direct implications for mining [110]. Small-to-intermediate mining companies must be based on equipment. Such activities constitute a significant proportion in terms of the number of transactions of the mining market.

Traditional technology in the mining industry is, in some cases, underserved. With continuous IoT advances and a decline in networking and sensor prices, profitable devices can be used for smaller operations.

Wireless technological innovations such as cellular network infrastructure with fifth generation (5G) technologies would promote the use of device-to-device (D2D) networks and reach wider areas and make digital capacities easier to access even remote and smaller mines. As secure communication networks for underground mining sites continue to be established, there will probably be progress in optimizing production and decision-making beyond short interval control and in dynamically optimizing the mining process. 5G connectivity will also reduce the risk of mining automation because the autonomous equipment is more familiar with its environment.

Academy Responsibilities in Mining DT Improvement

Universities and research centers can support the mining industry to have more benefit from DT in different ways. Data science and information technology methods have potential applications, particularly in the future mining engineers and specialists, in all the fields of study defined.

The eleven fields relevant to academic research or curriculum development in DT were presented in Table 1.4 and formed as a summary of the digital initiatives and current literature. Table 1.4 is not comprehensive; it provides enough guidance to concentrate on the future DT research in the mining industry.

Summary

DT is the technology-driven manner in which companies are increasingly adjusting to a digital world. Practical DT strategies include the management of information, communication, and decision-making. In comparative digitalization, the mining industry is behind many other industries. Prestigious multinational mining companies now focus digitally, setting the standard for smaller businesses. In the higher education industry, advanced data analytics systems are growing, and mining companies are finding digitally educated workers with great difficulty. This means that the mining engineering curriculum is questioned and a catalyst for the development of new fields of research. Universities and research centers are in a position to provide the DT for the existing mining companies, such as maturity development, education, supply chain management, health and accountability, lower impacts on the environment, operational excellence, quality of innovation, increasing challenges to innovation, alternative investment, and business intelligence.

TABLE 1.4

DT Research Areas and Their Tools Are Known

Application	Explanation
Personnel development	Generate a collaborative and an innovative workplace culture [14,66,88,89,103,104,111]
Training	Offer digital expertise through a transition in working habits to the existing and future workforce [27,101,112]
Control of the supply chain	Develop element- and product-tracking techniques [68,91,113,114] Remove conflict smuggling and minerals [10,109,110] Advance global market interaction [12,37,41,42,66,104]
Safety	Enhance and protect workers against damage and danger [64,70,85,115,116]
Clarity	Improving digital content, consistency in data quality, accessibility, confidence, autonomy, and environmental enforcement and monitoring infrastructure [10,117–119]
Reduction of environmental impacts	Supervise tailings and accelerate the reclamation procedure [51,52,61,120–127]
Operational excellence	Create and refine market simulations based on data [128–132]
Standards for creativity	Forms to creative technology laws and technical best practice [117,126,133]
Reduce barriers to creativity	Complete the Innovation Gap in the development of new technology commercialization plans [6,17,32,38,97,101,134]
Many savings	Enhance ties with stakeholders and build new alliances [11,111,133]
Intelligence for business	Provide decision-makers with insights and help [41,43,44,72,79,104]

References

1. Trist, E.L. and K.W. Bamforth, Some social and psychological consequences of the longwall method of coal-getting: An examination of the psychological situation and defences of a work group in relation to the social structure and technological content of the work system. *Human Relations*, 1951. 4(1): pp. 3–38.
2. Cooper, R. and M. Foster, Sociotechnical systems. *American Psychologist*, 1971. 26(5): pp. 467.
3. Westerman, G., D. Bonnet, and A. McAfee, *Leading Digital: Turning Technology into Business Transformation*. 2014: Harvard Business Press, Boston, MA.
4. Soofastaei, A., The application of artificial intelligence to reduce greenhouse gas emissions in the mining industry. In *Green Technologies to Improve the Environment on Earth*. 2018, IntechOpen.
5. Soofastaei, A., Introductory chapter: Advanced analytics and artificial intelligence applications. In *Advanced Analytics and Artificial Intelligence Applications*. 2019, IntechOpen.
6. Hinings, B., T. Gegenhuber, and R. Greenwood, Digital innovation and transformation: An institutional perspective. *Information and Organization*, 2018. 28(1): pp. 52–61.
7. Power, D.J. and C. Heavin, *Data-based Decision Making and Digital Transformation*. 2018: Business Expert Press, New York.

8. Sousa, M.J. and Á. Rocha, Digital learning: Developing skills for digital transformation of organizations. *Future Generation Computer Systems*, 2019. 91: pp. 327–334.

9. Ebert, C. and C.H.C. Duarte, Digital transformation. *IEEE Software*, 2018. 35(4): pp. 16–21.

10. Troshani, I., et al., Digital transformation of business-to-government reporting: An institutional work perspective. *International Journal of Accounting Information Systems*, 2018. 31: pp. 17–36.

11. Gale, M. and C. Aarons, Digital transformation: Delivering on the promise. *Leader to Leader*, 2018. 2018(90): pp. 30–36.

12. von Leipzig, T., et al., Initialising customer-orientated digital transformation in enterprises. *Procedia Manufacturing*, 2017. 8: pp. 517–524.

13. Vial, G., Understanding digital transformation: A review and a research agenda. *The Journal of Strategic Information Systems*, 2019. 28: pp. 118–144.

14. Mauri, P. Understanding the relationships between organizations and information technologies. The Role of Mapping. in STPIS@ CAiSE. 2018.

15. Atzori, L., A. Iera, and G. Morabito, The internet of things: A survey. *Computer Networks*, 2010. 54(15): pp. 2787–2805.

16. Gandhi, P., S. Khanna, and S. Ramaswamy, Which industries are the most digital (and why). *Harvard Business Review*, 2016 (1): pp. 45–48.

17. Sganzerla, C., C. Seixas, and A. Conti, Disruptive innovation in digital mining. *Procedia Engineering*, 2016. 138: pp. 64–71.

18. Young, A. and P. Rogers, A review of digital transformation in mining. *Mining, Metallurgy & Exploration*, 2019. 36: pp. 1–17.

19. Yanzhou Coal Mining Company Ltd., Annual Report. 2018. Yanzhou Coal Mining Company Ltd., Zoucheng City.pp. 23–31.

20. China Shenhua Energy Company Ltd., Annual Report. 2018. China Shenhua Energy Company Ltd., Beijing. pp. 185–201.

21. First Quantum Minerals Ltd., Annual Report. 2018. First Quantum Minerals Ltd., Toronto. pp. 11–19.

22. Franco-Nevada Corporation, Annual Report. 2018. Franco-Nevada Corporation, Toronto. pp. 86–93.

23. Freeport-McMoRan, Annual Report. 2018. Freeport-McMoRan, Phoenix, AZ. pp. 23–34.

24. Jiangxi Copper Company Ltd., Annual Report. 2018. Jiangxi Copper Company Ltd., Guixi City. pp. 44–52.

25. Gale, M., *The Digital Helix: Transforming Your Organization's DNA to Thrive in the Digital Age*. 2017: Greenleaf Book Group, Austin, TX.

26. Setia, P., et al., Leveraging digital technologies: How information quality leads to localized capabilities and customer service performance. *Mis Quarterly*, 2013. 37: pp. 565–590.

27. Cooper, J. and K.D. Weaver, *Gender and Computers: Understanding the Digital Divide*. 2003: Psychology Press, Philadelphia, PA.

28. Kajee, L., Digital literacy: A critical framework for digital literacy practices in classrooms. In *EDULEARN16 Proceedings*. https://doi. org/10.21125/edulearn, 2016.

29. Martin, A., Digital literacy and the "digital society". *Digital Literacies: Concepts, Policies and Practices*, 2008. 30: pp. 151–176.

30. Kowalski-Trakofler, K.M., L.J. Steiner, and D.J. Schwerha, Safety considerations for the aging workforce. *Safety Science*, 2005. 43(10): pp. 779–793.

31. Prevention, C.f.D.C.a Mining Facts. 2015 [cited 2019; https://www.cdc.gov/niosh/mining/works/statistics/factsheets/miningfacts2015.html].

32. Cory, S., Australia: Investing in innovation. *American Association for the Advancement of Science*, 2001. 293: p. 2169.

33. Kane, G.C., et al., Strategy, not technology, drives digital transformation. *MIT Sloan Management Review and Deloitte University Press*, 2015. 14: pp. 1–25.

34. Gren, L., A. Wong, and E. Kristoffersson, Choosing agile or plan-driven enterprise resource planning (ERP) implementations--A study on 21 implementations from 20 companies. arXiv preprint arXiv:1906.05220, 2019.

35. Ganesh, K., et al., *Enterprise Resource Planning: Fundamentals of Design and Implementation*. 2014: Springer, Berlin.

36. Stubbs, E., *The Value of Business Analytics: Identifying the Path to Profitability*. Vol. 43. 2011: John Wiley & Sons, San Francisco, CA.

37. Broekhuizen, T.L., T. Bakker, and T.J. Postma, Implementing new business models: What challenges lie ahead? *Business Horizons*, 2018. 61(4): pp. 555–566.

38. Lee, J. and K. Prowse, Mining & Metals+Internet of Things: Industry opportunities and innovation. MaRS Discovery District market report. 2014. http://www.marsdd.com/news-and-insights/mining-industry-iottechnology.

39. Hemphill, T.A. and G.O. White III, The world economic forum and nike: Emerging 'Shared Responsibility'and institutional control models for achieving a socially responsible global supply chain? *Business and Human Rights Journal*, 2016. 1(2): pp. 307–313.

40. Fryman, L., G. Lampshire, and D. Meers, *The Data and Analytics Playbook: Proven Methods for Governed Data and Analytic Quality*. 2016: Morgan Kaufmann, Burlington, MA.

41. Isson, J.-P., and J. Harriott, *Win with Advanced Business Analytics: Creating Business Value from your Data*. 2012: John Wiley & Sons, Hoboken, NJ.

42. Serrano-Morales, C.A., C.-A. Berlioz-Matignon, and F. Mangin, Rule-based management of adaptive models and agents. 2012, Google Patents.

43. Chen, H., R.H. Chiang, and V.C. Storey, Business intelligence and analytics: From Big data to big impact. *MIS Quarterly*, 2012. 36(4): pp. 1165–1188.

44. Minelli, M., M. Chambers, and A. Dhiraj, *Big Data, Big Analytics: Emerging Business Intelligence and Analytic Trends for Today's Businesses*. Vol. 578. 2013: John Wiley & Sons, Hoboken, NJ.

45. Ohlhorst, F.J., *Big Data Analytics: Turning Big Data into Big Money*. Vol. 65. 2012: John Wiley & Sons, Hoboken, NJ.

46. Dasgupta, N., *Practical Big Data Analytics: Hands-on Techniques to Implement Enterprise Analytics and Machine Learning Using Hadoop, Spark, NoSQL and R*. 2018: Packt Publishing Ltd., Birmingham.

47. Fekete, J.A., Big data in mining operations. 2015, Masters thesis, Copenhagen Business School, University of Copenhagen, Denmark.

48. Gilchrist, A., *Industry 4.0: The Industrial Internet of Things*. 2016: Apress, New York.

49. Shneiderman, B. and P. Scheuermann, Structured data structures. *Communications of the ACM*, 1974. 17(10): pp. 566–574.

50. Hobbs, L., et al., *Oracle 10g Data Warehousing*. 2011: Elsevier, Amsterdam.

51. Rogers, W.P., M.M. Kahraman, and S. Dessureault, Exploring the value of using data: A case study of continuous improvement through data warehousing. *International Journal of Mining, Reclamation and Environment*, 2019. 33(4): pp. 286–296.

52. Erkayaoglu, M. and S. Dessureault, Improving mine-to-mill by data warehousing and data mining. *International Journal of Mining, Reclamation and Environment,* 2019. 33(6): pp. 409–424.
53. Kennedy, J., Field experience with process data managers. Preprints-society of mining engineers of AIME. 1994.
54. La Rosa, D., The development of an information management system for the improvement of drilling and blasting in mining operations. In *Proceedings of 29th International Symposium Computer Applications in the Minerals Industries,* Beijing. 2001.
55. Wellman, T.A., Fleet management system. 2011, Google Patents.
56. Moradi Afrapoli, A. and H. Askari-Nasab, Mining fleet management systems: A review of models and algorithms. *International Journal of Mining, Reclamation and Environment,* 2019. 33(1): pp. 42–60.
57. Raab, D.M., Marketing systems for online media. *Information Management,* 2010. 20(2): p. 34.
58. Desai, N., IT vs. OT for the Industrial Internet-two sides of the same coin. GlobalSign Blog. 2016.
59. Hickson, D.J., D.S. Pugh, and D.C. Pheysey, Operations technology and organization structure: An empirical reappraisal. *Administrative Science Quarterly,* 1969. 14: pp. 378–397.
60. Wyatt, E.J., et al., An overview of the beacon monitor operations technology. *Network (DSN),* 1997. 1: p. 2.
61. Boulter, A. and R. Hall, Wireless network requirements for the successful implementation of automation and other innovative technologies in open-pit mining. *International Journal of Mining, Reclamation and Environment,* 2015. 29(5): pp. 368–379.
62. Caldwell, T., Plugging IT/OT vulnerabilities–part 1. *Network Security,* 2018. 2018(8): pp. 9–14.
63. Pires, J.N., *Industrial Robots Programming: Building Applications for the Factories of the Future.* 2007: Springer Science & Business Media, Boston, MA.
64. Sun, E., A. Nieto, and Q. Li. The drive fatigue pattern monitor for haul truck drivers in surface mining operations. In *2015 12th International Conference on Fuzzy Systems and Knowledge Discovery (FSKD).* Zhangjiajie, 15–17 August, 2015. IEEE.
65. Notley, S.R., A.D. Flouris, and G.P. Kenny, On the use of wearable physiological monitors to assess heat strain during occupational heat stress. *Applied Physiology, Nutrition, and Metabolism,* 2018. 43(9): pp. 869–881.
66. World-Economic-Forum. World Economic Forum White Paper. Digital Transformation of Industries: In collaboration with Accenture. 2016. World Economic Forum White Paper. Digital Transformation of Industries.
67. Chaulya, S. and G. Prasad, *Sensing and Monitoring Technologies for Mines and Hazardous Areas: Monitoring and Prediction Technologies.* 2016: Elsevier, Amsterdam.
68. Job, A. and P.R. McAree, Three case studies on the implementation of new technology in the mining industry. 2017.
69. Balaba, B., M.Y. Ibrahim, and I. Gunawan. Utilisation of data mining in mining industry: improvement of the shearer loader productivity in underground mines. In *IEEE 10th International Conference on Industrial Informatics.* Beijing, July 25–27, 2012.

70. Singh, A., D. Kumar, and J. Hötzel, IoT based information and communication system for enhancing underground mines safety and productivity: Genesis, taxonomy and open issues. *Ad Hoc Networks*, 2018. 78: pp. 115–129.
71. Jeschke, S., et al., Industrial internet of things and cyber manufacturing systems. In Jeschke, S., Brecher, C., Song, H., Rawat, D. (eds.) *Industrial Internet of Things*. 2017: Springer, Cham. pp. 3–19.
72. Jarrahi, M.H., Artificial intelligence and the future of work: Human-AI symbiosis in organizational decision making. *Business Horizons*, 2018. 61(4): pp. 577–586.
73. Kazakidis, V., Z. Mayer, and M. Scoble, Decision making using the analytic hierarchy process in mining engineering. *Mining Technology*, 2004. 113(1): pp. 30–42.
74. Watson, H.J., Preparing for the cognitive generation of decision support. *MIS Quarterly Executive*, 2017. 16(3): pp. 153–169.
75. Newell, S. and M. Marabelli, Strategic opportunities (and challenges) of algorithmic decision-making: A call for action on the long-term societal effects of 'datification'. *The Journal of Strategic Information Systems*, 2015. 24(1): pp. 3–14.
76. Zhou, C., et al., Industrial internet of things:(IIoT) applications in underground coal mines. *Mining Engineering*, 2017. 69(12): pp. 50.
77. Botto, F., *Dictionary of e-Business: A Definitive Guide to Technology and Business Terms*, 2ndEdition. ISBN: 978-0-470-84470-0 2003.
78. Subashini, S. and V. Kavitha, A survey on security issues in service delivery models of cloud computing. *Journal of Network and Computer Applications*, 2011. 34(1): pp. 1–11.
79. Erkayaoğlu, M., *A Data Driven Mine-To-Mill Framework For Modern Mines*. 2015: University of Arizona, Tucson, AZ.
80. Niosh, U., Workplace Safety & Health Topics. 2015, Hierarchy of Controls.
81. Iyer, B. and M. Subramaniam, The strategic value of apis. *Harvard Business Review*, 2015. 1(7): p. 2015.
82. Marvel, M.R., *Encyclopedia of New Venture Management*. 2012: Sage, London.
83. Zhang, P. and D.F. Galletta, *Human-Computer Interaction and Management Information Systems: Foundations*. Vol. 5. 2006: ME Sharpe, Armonk, NY.
84. Stahl, P., B. Donmez, and G. Jamieson. A field study of haul truck operations in open pit mines. In *Proceedings of the Human Factors and Ergonomics Society Annual Meeting*. 2011. SAGE Publications: Los Angeles, CA.
85. Baldemair, R., et al., Evolving wireless communications: Addressing the challenges and expectations of the future. *IEEE Vehicular Technology Magazine*, 2013. 8(1): pp. 24–30.
86. Gupta, A. and R.K. Jha, A survey of 5G network: Architecture and emerging technologies. *IEEE Access*, 2015. 3: pp. 1206–1232.
87. Hoover, H. and L.H. Hoover, *De Re Metallica*. 1950: Courier Corporation, Chelmsford, MA.
88. Aggarwal, I. and A.W. Woolley, Team creativity, cognition, and cognitive style diversity. *Management Science*, 2019. 65(4): pp. 1586–1599.
89. Huo, D., K. Motohashi, and H. Gong, Team diversity as dissimilarity and variety in organizational innovation. *Research Policy*, 2019. 48(6): pp. 1564–1572.
90. McMahon, T., The value of process automation. *Chemical Engineering Progress*, 2008. 104(3): pp. 28–28.
91. Souza, G.C., Supply chain analytics. *Business Horizons*, 2014. 57(5): pp. 595–605.

92. Provost, F. and T. Fawcett, *Data Science for Business: What You Need to Know about Data Mining and Data-Analytic Thinking*. 2013: O'Reilly Media, Inc., Newton, MA.

93. Knobel, M., *Digital Literacies: Concepts, Policies and Practices*. Vol. 30. 2008: Peter Lang, New York.

94. Tylor, Data workers prove a rare resource for Barrick. 2017. pp. 8–10.

95. Strharsky, J., The future of mining: More digging through data than strata. *AusIMM Bulletin*, 2017. 2017: p. 18.

96. Kubacki, T., et al., Changes in mining-induced seismicity before and after the 2007 Crandall Canyon Mine collapse. *Journal of Geophysical Research: Solid Earth*, 2014. 119(6): pp. 4876–4889.

97. Galindo-Martín, M.-Á., M.-S. Castaño-Martínez, and M.-T. Méndez-Picazo, Digital transformation, digital dividends and entrepreneurship: A quantitative analysis. *Journal of Business Research*, 2019. 101: pp. 522–527.

98. Westerman, G. and D. Bonnet, Revamping your business through digital transformation. *MIT Sloan Management Review*, 2015. 56(3): p. 10.

99. Joyce, S., Major issues in miner health. *Environmental Health Perspectives*, 1998. 106(11): pp. A538–A543.

100. Heine, C. and D. O'Connor, *Teaching Information Fluency: How to Teach Students to be Efficient, Ethical, and Critical Information Consumers*. 2013: Scarecrow Press, Lanham, MD.

101. McCarthy, F. and M. Vickers, Digital natives, dropouts and refugees: Educational challenges for innovative cities. *Innovation*, 2008. 10(2–3): pp. 257–268.

102. Yoo, Y., A. Bryant, and R.T. Wigand, Designing digital communities that transform urban life: Introduction to the special section on digital cities. *CAIS*, 2010. 27: p. 33.

103. Holtel, S., Artificial intelligence creates a wicked problem for the enterprise. *Procedia Computer Science*, 2016. 99: pp. 171–180.

104. Dirican, C., The impacts of robotics, artificial intelligence on business and economics. *Procedia-Social and Behavioral Sciences*, 2015. 195: pp. 564–573.

105. Blom, M., A.R. Pearce, and P.J. Stuckey, Short-term planning for open pit mines: A review. *International Journal of Mining, Reclamation and Environment*, 2019. 33(5): pp. 318–339.

106. Beck, K., et al., Manifesto for Agile Software Development. 2001.

107. Cooke, J.L., *Agile Productivity Unleashed: Proven Approaches for Achieving Real Productivity Gains in Any Organization*. 2014: IT Governance Publishing, Ely.

108. Koller, S., et al., Using virtual reality for forensic examinations of injuries. *Forensic Science International*, 2019. 295: pp. 30–35.

109. Roberts, J., Blockchain in real life. *Fortune*, 2017. 176(3): p. 49.

110. Chen, Y., Blockchain tokens and the potential democratization of entrepreneurship and innovation. *Business Horizons*, 2018. 61(4): pp. 567–575.

111. Cutifani, M. and P. Bryant, Reinventing mining: Creating sustainable value. Kellogg Innovation Network. 2015.

112. Meyen, E., Significant advancements in technology to improve instruction for all students: Including those with disabilities. *Remedial and Special Education*, 2015. 36(2): pp. 67–71.

113. Diaz, R., et al. Diagnosis of process health, its treatment and improvement to maximise plant throughput at goldfields Cerro Corona. In *SAG 2015 Conference Proceedings*. Vancouver. 2015.

114. Isokangas, E., et al., Using SmartTag to track ore in process integration and optimization projects: Some case studies in a variety of applications. In *Platinum 2012, SAIMM Conference*. Sun City, North West, 18–21 September, 2012.

115. Thrybom, L., et al., Future challenges of positioning in underground mines. *IFAC-PapersOnLine*, 2015. 48(10): pp. 222–226.

116. Sun, E., et al., An integrated information technology assisted driving system to improve mine trucks-related safety. *Safety Science*, 2010. 48(10): pp. 1490–1497.

117. Preuveneers, D., W. Joosen, and E. Ilie-Zudor. Data protection compliance regulations and implications for smart factories of the future. In *2016 12th International Conference on Intelligent Environments (IE)*. London, 14–16 September, 2016.

118. Augusto, J.C., et al., Intelligent environments: A manifesto. *Human-Centric Computing and Information Sciences*, 2013. 3(1): pp. 1–18.

119. Lavrin, A. and M. Zelko, Moving toward the digital factory in raw material resources area. *Acta Montanistica Slovaca*, 2010. 15(3): p. 225.

120. Bamford, T., K. Esmaeili, and A.P. Schoellig, A real-time analysis of post-blast rock fragmentation using UAV technology. *International Journal of Mining, Reclamation and Environment*, 2017. 31(6): pp. 439–456.

121. Benito, R. and S. Dessureault, Estimation of incremental haulage costs by mining historical data and their influence in the final pit limit definition. *Mining Engineering*, 2008. 60(10): pp. 44–49.

122. Duffy, K., et al., Resource efficient mining processes of tomorrow. In *Proceedings of Future Mining Conference*. Sydney, Australia. 2015.

123. Córdova, F., et al., A proposal of logistic services innovation strategy for a mining company. *Journal of Technology Management & Innovation*, 2012. 7(1): pp. 175–185.

124. Pan, X., System integration of automated mine optimization system. *IFAC Proceedings Volumes*, 2013. 46(16): pp. 148–154.

125. Li, Q.-M., H. Zhang, and Z. Yang. Digital tailings system for non-coal mine solid waste safety treatment. In *2017 3rd International Forum on Energy, Environment Science and Materials (IFEESM 2017)*. Shenzhen, November 25–26, 2017.

126. Muñoz, J.I., R.R. Guzmán, and J.A. Botín, Development of a methodology that integrates environmental and social attributes in the ore resource evaluation and mine planning. *International Journal of Mining and Mineral Engineering*, 2014. 5(1): pp. 38–58.

127. Soofastaei, A., et al., Development of a multi-layer perceptron artificial neural network model to determine haul trucks energy consumption. *International Journal of Mining Science and Technology*, 2016. 26(2): pp. 285–293.

128. Valery, W., A. Jankovic, and B. Sonmez. New methodology to improve productivity of mining operations. In *Proceedings of XIV Balkan Mineral Processing Congress*, Turkey. 2011.

129. Lala, A., et al., Productivity in mining operations: Reversing the downward trend. *AusIMM Bulletin*, 2016. 2016: p. 46.

130. Iansiti, M. and R. Levien, *The Keystone Advantage: What the New Dynamics of Business Ecosystems Mean for Strategy, Innovation, and Sustainability*. 2004: Harvard Business Press, Boston, MA.

131. McNab, K. and M. Garcia-Vasquez, Autonomous and remote operation technologies in Australian mining. Prepared for CSIRO Minerals Down Under Flagship, Minerals Futures Cluster Collaboration, by the Centre for Social

Responsibility in Mining, Sustainable Minerals Institute, The University of Queensland. Brisbane, 2011.

132. Soofastaei, A., Introductory chapter: Advanced analytics and artificial intelligence applications. In Soofastaei, A. (ed.) *Advanced Analytics and Artificial Intelligence Applications*. 2019: InTechOpen, London. pp. 10–22.

133. Moffat, K., et al., The social licence to operate: A critical review. *Forestry: An International Journal of Forest Research*, 2016. 89(5): pp. 477–488.

134. Yanzhou Coal Mining Company, M., Annual Report. 2017. Yanzhou Coal Mining Company: Zoucheng City. pp. 22–27.

2

Advanced Data Analytics

Ali Soofastaei

Introduction

The digital age with its possibilities and uncertainty confounds industries and economies, with substantial possible knowledge in each section. A new model for the company in this time is being generated by the tendency toward the data lake and drawing upon the hidden knowledge. The influence of information leads companies to be agile and to achieve their objectives. Big data analytics (BDA) allows companies to recognize, evaluate, anticipate, and administer covert opportunities for growth to attain commercial value [1]. In order to generate information from that data, BDA uses advanced analysis techniques that affect the decision-making process to reduce the complexity of the process [2]. In order for BDA to process, analyze, and perform highly precise analysis in real time, it needs a new and advanced algorithm. Computer and deep learning (DL) use this tool to distribute their complex algorithms and consider the problematic approach [3].

A vast amount of data, profound learning and algorithms, machine learning (ML), and similar approaches have been included in this investigation. This offers a theoretical model for the algorithm's relationship, which facilitates the implementation of IoT data for BDA by researchers and practitioners.

Figure 2.1 illustrates the process of thinking about DL and ML approaches.

Big Data

The creation of a large number of raw data is one of the major implications of the digital world. The manager's role is concerned with the distribution of such important resources in various forms and sizes based on the needs of the organizations. Big data have the ability in every part of society to influence the social aspects of education. The subject of raw data management is becoming far more important as data volumes increase, in particular, in the

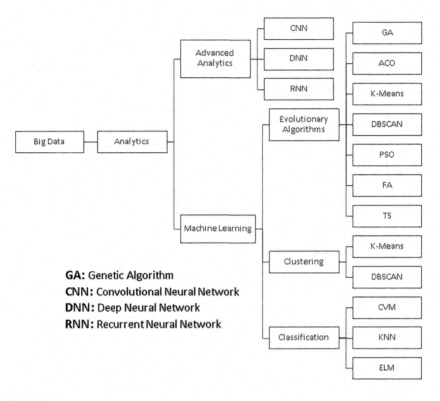

FIGURE 2.1
Big data analysis approaches.

technology-based companies. Through contrast, advanced techniques can overcome their complexity, through comparison with raw data features such as variety, velocity, and the volume of big data. Therefore, for "experiment," "simulation," "data analysis," "monitoring," BDA were suggested. As one of the BDA approaches, ML provides a system that allows predictive analysis dependent on supervised and unattended data input. In mutual partnerships, the influence of computer science analysis and data entry exists. The better and more reliable suggestions, the analyses will be. DL is also used as an input of the machine to learn from secret data patterns [3].

Analytics

Big data have been introduced in the digital era of increasing data consumption levels, known for their large volume, variety, veracity, speed, and reliability. This gives rise to the uncertainty and massively of different data forms: structured, semistructured, and nonstructured. It entitles the company to use new approaches and tools in analytical aspects. Big data analysis is, therefore, regarded as a sophisticated method to tackle large volume complexity [4]. Chen Chiang (2012) first coined large data analytics, demonstrating the relationship between business intelligence and science and creating a secure connection to data mining and statistical analysis [5].

BDA encourages innovation productivity and performance organizations [6]. BDA has been described as methods that can detect cached models and offer insights into interesting relationships through analysis, treatment, discovery, and exhibition of the results in understudies contexts [7]. BDA has the benefit of reducing the complexity and control of cognitive pressures in the information society. Furthermore, feature identify is the most important feature leading BDA to success. It involves the description of the main characteristics that have a major impact on performance. It is accompanied by the identification of correlations between input and dynamic levels [7].

Due to the rapid growth of the e-business and deep integration of BDA, the global economy has grown stronger. Governments can use BDA to represent their people better [7].

Data can be handled and analyzed using BDA as a special application in a business environment. Big data can easily be used by social networks to analyze large data. It allows users to understand the customer's conduct and to handle five big data features that are identified as length, size, value, diversity, and truthfulness. BDA not only helps businesses to develop a holistic understanding of customer behavior but allows firms to adopt new strategies [8] further. BDA is used by medium and small firms to collect semistructured data, which results in improved quality products and website design [8]. BDA benefits from using large data to improve firm efficiency with the application of technology and technology [9].

Intuition is increasing the importance of BDA, a result of the incorporation of different information in the decision-making process [8]. This turns the decision-making process into a field based on evidence. The detailed removal of big data was split into two main methods: data management and data analysis; first, technical support for data collection, storage, and study planning. Therefore, the large-scale data analysis is regarded as a subprocess of insight. The key methods for analyzing text, audio, video, and social media data were included. BDA can be concluded as the main tool for digital information collection and understanding [10]. Data storage, data management, data analysis, and data visualization also are included [9].

Big data analysis can build an efficient and effective value in the organization's organizational and strategic strategies and plays a role as a game-change tool in increasing productivity [11].

Industry professionals agree that the next "blue ocean" for big data analysis provides companies with opportunities [12], and it is called the "fourth scientific model" [13].

In order to deal with BDA, ML and DL have been established. Various areas such as "medicine" and "Internet of Things" (IoT) use ML to test the predictive capabilities for big data and Search Engines. In other words, learning patterns are generalized for predicting the future outcomes. Two main ML components are the feature design and data representation. Also, useful big data extraction is why DL is deployed, a technology inspired by the human brain to process neural signals as a subproduction of ML [3].

Deep Learning

DL was launched in the 1940s [14]. In 2006, however, the birth of DL algorithms was decided when Hinton implemented a layer-wise-greedy-learning approach to resolve the limitation of the neural network (NN) system by trapping in the local optima point to find optimized points. Unregulated learning before layer training is the idea behind the proposed Hinton Process [15].

DL algorithms remove complex and highly abstract hidden features. When volumes of unstructured data are represented, the layered DL algorithm architecture works effectively. The goal of DL is to introduce multiple transition layers, which will display each layer [16]. BDA includes the entire knowhow that has been developed by an in-depth research. The main feature of BDA is the extraction of underlying features by a large quantity of data [16].

DL is an ML subfield implemented when conditions have been developed, such as increased chip-processing, which leads to vast amounts of data, lowering computer hardware costs, and remarkable creation of ML algorithms [17]: (1) Sparse Coding, (2) Constricted Boltzmann Robots, (3) Autoencoder, and (4) Convolutional Neural Networks (CNN).

CNNs

CNN has a DL algorithm of the convolutional and subsample layer design based on the NN model. In case the value of the instances is every data point, multiple information is used as a container [18].

Three functions, namely, local layer, subsampling, and weight sharing, were known to CNN. The three layers are composed of the input layer, the surface layer, and the output layer of the sample. Every convolution layer in the hidden layer is after the subsample layer. CNN training is performed in two phases of feeding development, during which the previous level results are entered, and the backpropagation is conducted through a reverse error process and a hierarchical process [19]. The nonlinear transformation method is used in a first-level convolution process to transform each case into a nonlinear space in various phases of filtering. Instead, in the max-pooling layer, which is representative of an instance container, the transformed nonlinear space in this step was taken by looking at every instance's maximum response in the filtering process. The picture gives the maximum answer to a large pie to determine the status of individual cases in each class. It results in a pattern of classification [18].

The CNN is an artificial learning process using data with two convolution and bundling layers, derived characteristics. The multilayer interpretation of the characteristics gained in the classification process is another aspect of CNN [20].

Deep Neural Network

A detailed computational architecture of the supervised data developed for computer algorithms and methods is called deep neural networks (DNNs) [20]. This comes from the shallow neural artificial networks (SANNs) linked to artificial intelligence (AI) [21].

DNN uses a dynamic layered architecture to handle complexity and multiple strata because the hierarchical architecture of the layers collection can provide nonlinear knowledge [20].

DNN is one of the leading classification methods [22] due to its exceptional implementation in the categorization of complex issues. One of the most difficult questions in DNN is its output as it attempts to reduce an objective function with a high-ranking number of factors in a multilingual search area in cases of optimization problems. Therefore, a reasonable DNN optimization algorithm is attracted and trained at a high level. DNN is made of a denoising denouncer (SDAE) system [23] which is a lot of cascades and SoftMax classifiers for autoencoder layers. The first uses raw data to produce new functionality, and SoftMax is used to perform the function classification process accurately. The features mentioned above are mutually complementary and help DNN achieve its central and effective classification performance. The gradient descent (GD) optimization algorithm may be used in linear problems, in particular not complex goals in DNN training, and its key condition is that the quantity of the optimization parameter is similar to the optimization solution [24].

Recurrent Neural Network (RNN)

Through the 1980s, RNN, the network of nodes like a neuron, was developed. Every neuron-like node can be interconnected and divided into input, secret, and output neuron categories. In this triple process, the data are received, transformed, and produced. The time-limited activation is a function of every neuron, and each synapse has a justifiable real weight [25]. In addition to learning and learning (Yang. F et al. 2016), a classifier for NNs performed extremely well in dynamic system modeling using the current information approach [26,27]. RNN is slightly different from the artificial NN from which it comes with the context of the brain algorithm inspired by human. The RNN Research focuses on a variety of fields, including the areas of image processing, signal processing, associative recollections, pattern recognition, robotics, and control [1]. RNN can view past knowledge comprehensively and apply it to sudden modifications through input and feedback. RNN also can access time-varied data to simplify the NN architecture. It suits well in real-time problems with its easiness and different features [28]. In the hierarchy system, RNN can process temporal data and take a multilayer of abstract data to show dynamic features [29]. RNN can link signals at different levels, bringing substantial processing power with a huge amount of space in the memory [30].

ML

ML was characterized by interpreting data, followed by learning algorithms as predictive algorithms in a nonstructured system. Three major ML categories are supervised, unsupervised, and reinforcement learning (see Figure 2.2) [31], which is done in the step of preprocessing data, learning, and evaluation.

Preprocessing is connected to turning the raw information into the correct structure, which can be implemented during a certain stage of learning, such as data purification, collecting, transforming, and mixing it. The information shall be carefully chosen, and the results, statistical assessments, and error estimation or deviation assessments shall be evaluated during the assessment process. This can change the selected learning process parameters [32]. The first is the analysis of features that are important to identify from some training data. For the evaluation of unscored data, the data were deployed and trained in the testing algorithm. After the unlabeled data is processed, the production will be generated that is able to be categorized if it is discrete or regressive.

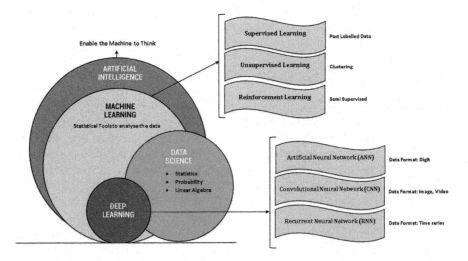

FIGURE 2.2
Artificial intelligence, machine learning, data science, and deep learning definition and integration.

On the opposite, ML can be used without a period of training in pattern recognition, which is called unregulated ML. In that category, additional ML forms, which is an organization, will be formed when patterns of features are used for grouping data, and if the secret data rules are recognized [33]. In other words, natural grouping from data unlabeled is the key mechanism of unsupervised ML or clustering. In this method, in many clusters compared which consider relationship measure, in a data set, the K cluster is very close. The "Overlapping," "Hierarchical," "Partitioned," techniques are the three types of unsupervised ML. Two types of hierarchical approaches are "agglomerative" and "divisive." The first is called an aspect that produces a separate cluster with the propensity to participate in a greater cluster; the second, however, is a greater cluster that is to be separated into less significant ones. The "partitioned" method begins with the development of multiple disjoint clusters out of data sets, with no hierarchical structure considered. In contrast, the "overlap" method is characterized as a way to find fluid or diffused partitions that have been carried out by "relaxation of mutually disjointed restrictions." "Simplicity" and "efficacy" are the two key unregulated technique characteristics [31].

Fuzzy Logic

In many areas, from engineering to data analysis, Lotfi Zadeh (1965) suggested fuzzy logic was used. ML also benefits from abstract logic as fluid, inductively inductive ML. The changes were made on the grounds of "fuzzy rule-making," "fuzzy decision-making structures," "substantial neighbor forecasts," or "fuzzy vector support machines" [34].

Classification Techniques

Classification is one of the critical aspects of ML [35]; i.e., the first step in the analysis of data [36]. New fields such as face detection and handwriting recognition have been identified previously in previous studies. The operating classification algorithm was split into two offline and online groups [35]. The static data set is used for training in an offline approach. After the training process is completed, the classifier will stop the training, and no modification of the ordered data will be permitted.

The type one pass which learns from various information is, however, specified by the online category. The key characteristics of the data are retained in the memory and processed until the data are removed. Two key strategies for the online group are the gradual and evolved processes of data change in an unstable environment resulting from an evolutionary system structure and constant updates of meta-parameters [35].

In 1995, Cortes and Vapnik proposed the Vector Machine (SVM) to solve the MD and regression problems as the excellent learning outcomes [37]. SVM generates a hyperplane dividing information into binary categories during this process; the main objective being to find the most significant margin in binary categories that takes hyperplane space into account [38]. Statistical learning theory is an integral part of the process of SVM growth, using the statistical learning method and the kernel method to achieve a sufficient number of students, because of the risk-reduction mechanism [39].

K-nearest neighbor algorithm uses object classification in the nearest training class and is one of the most common algorithms in data mining and extraction classifying issues. The closest neighbors are allocated an entity. It works like that. The quality of this technology depends on the weighted qualifications. Some of the problems with this approach are as follows:

- Depending heavily on the K value, which is a measure of how the area is calculated.
- The method lacks the capacity for the distinction between near and distant neighbors.
- In the vicinity of neighbors, overlap or noise may occur [40].

KNN was first introduced as one of the most important data-mining algorithms for classification problems that are broadened to model recognition and ML technology. Expert programs are using KNN classification problems. Three primary KNN classificatory, which focus on k in any test sample class, are:

- "Nearest neighbor classifier (LMKNN) local mean-based:" Even though this approach can overcome a real outer negative influence, LMKNN is likely to be misclassified because the single value of k takes into account the scale of the district per class and applies this in all classes.
- "The nearest neighbor classifier (LMPNN) local medium-based pseudo-classifier:" The LMKNN and PNN methods are planned, in addition to esthetics. We are considered to be a strong multimodal median vector based on the nearest and pseudoneighbor multilocal median vectors for each class. Nevertheless, the weight of all classes cannot be generally understood as separate data from the next classification set [41].
- "Classification multilocal k-harmonic (MLMKHNN) nearest neighbor:" For classification decision law, MLMKHNN as an addition to KNN will take the middle harmonic range. It uses multilocally mean vectors of the closest neighbors by the class of all query samples, and the result of this phase will be a harmonic mean distance [42]. Such approaches are designed to find different decisions about classification [41].

In 2006, Huang et al. introduced a classification system for an extreme learning machine (ELM), which works in a NN using the hidden single-layer feedback. The weight input and deviation are generated randomly in this layer, and the lowest square method is used to measure the weight of the product [36], which distinguishes from the traditional methods of this method. This is where learning takes place, and then, the matrix of transformation is located. It is used to reduce the number of error squares. The product of a feature reduction is then used for classification or dimensional reduction [43]. NNs are broken down into two different groups, neural and feedback networks, while ELM is one of the first, with a strong ability to learn, particularly in the resolution of highly complex nonlinear functions. In addition to fast learning methods, ELM utilized this function to solve common NN problems without iteration at a higher speed compared to the traditional NN in mathematical transition [44].

Given the efficiency of ELM in classification issues, the lack of ELM in binary classification issues is evident as an equivalent training stage in ELM is required in these problems. The problem is solved by the Twin Extreme Learning Machine classification simultaneous train and two nonparallel classification hypercrafts. Growing hyperplane is entered into a minimization function to reduce the distance with one class away from other classes [45]. ELM concentrates work in the description of data streams [46].

Clustering

As a guided learning process, clustering is intended to create groups of clusters that are similar and different from one another in terms of features [47]. In relation to its distance from one point in the other clusters, the distance between each observation is small [48]. The main uses for the clustering methods are "grouping," "explorative model-analysis," "decision-making," and "machine-learning situations." Four cluster groups are "hierarchical clustering," "clustering partitioning," "clustering based on the scale," and "clustering based on modeling" [49]. The problem of clustering is divided into two classes of generative and discriminatory approaches. First of all, the probability of sampling used to learn from the generated models is maximized, and second, similarities used on a pair basis that maximizes the similarities between the intercluster groups and minimizes the similarities between clusters [50].

The clusters of K-mean, Kernel K, spectral clustering, and clustering algorithms are available that have been studied for decades. After the introduction of K-means, k-means and spectral clusters establish a relation between the data and the functional space. Kernel and graph model are, respectively, implemented to obtain space in kernel k-means and spectral clusters. Also, spectral clustering uses the techniques of self-composition [51]. K-means, the clustering of multidimensional, multidimensional numerical data [52].

DBSCAN reflects clustering based on distance; it is typically a distribution field that is separate from the data set. This approach does not use a cluster a priori for the identification of clusters in the data. The user-defined parameter is considered to build clusters that are somewhat different from the parameter described in the clustering process [47].

Evolutionary Techniques

An optimal solution among several alternatives is the main objective of the problems in optimization. If the search area is that, it is difficult to provide the best solution. Specific strategies are needed to find the best solution, but the heuristic algorithm needs the best solution. Nonetheless, to overcome the deficiency, the best alternative was the population-based algorithm [53].

Genetic Algorithms (GAs)

GA is well-defined as the random search that aims in the complex, high-dimensional environment finding an almost optimal solution. The string of chromosomes is the key parameter for space searches to be carried out at their base and is called population, and a number of them tend to be a set. Following creating a random population, the best fitness and objective of each string are represented. The product of this step is inserted into the mattress, which is a few chosen strings with many copies. A new string generation will be produced from the string by deploying in the cross-over and mutation phase. The procedure is continued until a termination condition has been created. The following are a few examples of the application areas for the gas algorithm: image-processing, the NN, and ML [54]. GA is based on algorithms of genetic and natural selection, inspired by nature [55].

GA algorithms are intended to achieve the optimal solution by considering the starting point; even GA can find a perfect grouping of clustering methods [54]. In the sense of the set of functions, filter and wrapper are two major GA approaches. First, by applying heuristic data properties, including correlation, to investigate the importance of functionality and, second, by using the ML algorithm to evaluate the quality of the GA solution [56]. The optimized local point in K-means is based on seed values, and the generated clusters are based on original seed values. The local point is configured in K-means. GA conducts K-mean algorithms to find almost optimal or suitable cluster searches for initial seed values and covers the absence of K-mean algorithms [57]. The retrieval of information from the database is another way for GA to help create a classification system and rules for mining associations [58].

The selection of functionalities is important in big data as usually it has many features that explain empirical principles and select the right number of preprocessing features because data mining is important. Feature selection is divided into two groups which rely on the wrapper approach used by the learning algorithm and are independent of the learning algorithms that use the filter approach. Nonetheless, the filter method is different from the analysis algorithm. The optimal feature set will rely on the learning algorithm that is one of the key difficulties in choosing the filter. A wrapper approach is the best solution to check all kinds of functionalities using a learning algorithm. The major difficulty is its complexity, which can be overcome by GA in the calculation as a learning algorithm [59].

Ant Colony Optimization (ACO)

As a population-based stochastic process, Dorigo proposed a colonial optimization technique in 1991 [60]. The approach is biologically derived from the real ant behavior in food-search models [61]. The process is to deposit a chemical substance, known as a pheromone, on the ground as the anti continues to search for food toward the food source. The pheromone level will increase as the distance between the diet and the nest will be shorter. The route with a greater number of pheromones is used by new agents in this method. Both ants obey the positive feedback over time and choose the shortest route signposted with the greatest pheromone number [62]. Ant-colonial optimization applications have been described as problems in traveling salesmen, scheduling, structural engineering, power engineering, image processing, clustering, routing optimization algorithm in recent research [63], data mining [64], robot path planning [65], and DL [66].

Many benefits of the optimization approach for ants are as follows:

- the application of this strategy with other algorithms is less complex,
- power to find a nearly perfect solution,
- gain the benefit of parallel delivery (e.g., intelligent search),
- better work to maximize swarm intelligence, and
- high speed and accuracy [63].

The released substance called pheromone triggers the clustering of the species between the optimally placed species. The clustering of the ant-colony is used in the grid to cluster data objects [67].

All solutions are built in a single iteration, the movement is enhanced through local search, and the material is modified [35]. The critical moves toward ACO are as follows:

- pheromone trail initialization;
- to create a solution, use a pheromone trail; and
- pheromone trail warnings.

The pheromone update measures the pheromone fraction evaporation and the emission of pheromones that show the level of fitness of the solution [68].

Ant-colony decision tree (ACDT) is a branch of decision-making to determine decisions that are established in the current algorithm but are generated as a nondeterminative algorithm in every execution. The theory of ACDT is a pheromone trace on the edge of the classical algorithm and heuristic components.

After the limitations for optimization of a one-layer colony were announced, the multilayered colony algorithm was introduced when a solution could be sought, because it takes too long to develop an entity with massive quantities. This specifies the maximum volume of an object through transactions and defines a rough set of membership functions that can be strengthened by reducing the area of the quest through a subsequent refining process. The search ranges are also different, given the costs. The answer from each level consists of a new level, which is included in the approach but has to be changed with a smaller search area [69,70]. Tsang and Kwong suggested clustering of ant colonies to detect anomalies [70,71].

Bee Colony Optimization (BCO)

BCO algorithm works to inspire the actions of sweet bees, which is commonly used to automate issues such as "Salesman Traveling Problem," "Web Hosting Center," "Vehicle Routing," and the list continues. In 2005, Karaboga aimed a simple, comfortable, and little-to-control artificial bee colony (ABC) algorithm in optimization paradigms. The algorithm is a simple one. "Face recognition," "high-dimensional gene expression," and "speech segment classification" are examples of features identified and optimized by ABC and ACO in an extensive search area. Three types of bees are deployed on ABC algorithms known as "Employed Bees (EBees)," "Onlooker Bees (OBees)," and "Scout Bees deployed." This is achieved by placing the source of food and then passing on nectar information to OBees in which their number is similar to many food sources. The number of EBees is equal. The data are used until the final amount to control the food source. Scouts are engaged in the search for new food sources in declining food sources. The quantity of nectar is a quality element [71,72].

This is a two-stage approach: one that explores new information by bees and one that is connected to the exchange of data considering new alternatives by beehives.

Thus, he begins a bee to explore, who wants to find a complete path to his journey. When he leaves the hive, other bees perform spontaneous dances, equipped with several other bees known as "the preferred way," contribute to the cycle and require a complete path previously found by his partner, who took the bee to its place. The process of moving from Node to Node continues until the final destination is reached. The node of bees with two arc fitness and heuristic distance variables was selected with a heuristic algorithm. Bees can choose the shortest distance [53]. In the BCO algorithm, the extraction and exploration cycles are considered to be two alpha and beta values, respectively [73].

Particle Swarm Optimization (PSO)

Inspired by biological organisms, PSO was produced, especially the capability of the categorized animal to function together in an area to find the required position. In 1995, Kennedy and Eberhart introduced the approach as a stochastic population algorithm, known for characteristics such as trying to find a global optimizing point, easily implemented by taking a small number of parameters into the cycle of adjustment. It takes advantage of a highly productive search algorithm to work in various fields and problems of optimization of analysis [74].

The search method is carried out in a real value search area to address a nonlinear optimization problem. Through this process, the endpoint is found, which is the ideal point, through an iterative search. In other words, a particular space modernized by particle experience, or the best neighboring area, includes each particle's multidimensional study, and the objective function evaluates the fitness value of each particle. In each iteration, the best solution is retrieved. Once the optimal solution with particles is found, the best or best solution is the local solution, and the optimum point is called the best or best among the next particles [75]. Any solution is known as a particle with different features like the current position and speed in this algorithm. By taking different weights, the balance between the local and global searches can be modified. The trade-off between global and local searches is one of the key success factors in PSOs [76]. Some areas for PSO implementation include artificial NNs, pattern recognition, and fuzzy regulation. This algorithm provides social interaction and communication methods, such as "bird flock and fish education," to facilitate the exchange of cultural knowledge between swarming particles [77].

Firefly algorithm (FA)

In 2008, Yang unveiled the FA. The FA's principal idea is to assume that every firefly is unisexually attracted to other fireflies irrespective of gender. The highlight is the main attraction for a lightning fire, which pushes the less light into a higher. The range is opposed to appeal and illumination. The luminosity of the firefly is determined by the exercise area [78]. The degree of solution goodness is increased with the increasing luminosity of the fireflies. A maximum attraction model is proposed to show that all flames are lighter, and fires are identical when an important number of fires attract brighter fitness-based fireflies. Thus, during the search process, the convergence rate will be strong.

FA was inspired by the fireflies' lightening function and was known as the swarm intelligence algorithm. In some cases, FA works better compared to GM (GA) and PSO. The work area of the FA includes "unitary participation," "energy conservation," and "complex networks" [79]. Fluctuations can occur when the light sources are attracted by a large number of fireflies, and the search time is required. The FA (NaFA) Village Attraction reveals that fireflies are only attracted by some of the lightest areas that previous neighbors have known [80].

Tabu Search Algorithm (TS)

Tabu research is a metaheuristic approach that was planned and made better by Glover and Laguna (1997), which is a local search that takes the chance to find a globally optimized solution using numerical algorithms. TSs are based on and enhance edge projections. A solution that can be used for combinatory optimization is the local heuristic search tool [81]. The methodology's search process is versatile as adaptive memory is necessary. During various iterations, the process is done. A solution is found in every iteration. The solution is reachable via "move." There is a better solution in every move that can be stopped if no better answer is found [82]. The expectation parameters are an integral part of TS' search procedures that do not take into account the forbidden TS methods. In every solution, the requirements of the objective are met. The solutions are, therefore, both workable and time-consuming. A tabu list (LT), a short-term past, is used to proceed to TS. The short memory preserves the recent move by removing the old memory when the memory is complete to the highest level [83].

TS's key idea is to move into solution space, which is still unexplored, which would be a way of avoiding a local solution. Thus, "tabu" campaigns that are new are banned and discourage previous approaches from visiting them. This shows that the technique offers high-quality results [84].

BDA and IoT

The IoT is designed to create an innovative environment by sensing, processing, communicating, and socializing behaviors. Once IoT sensors have produced large raw data to process and analyze, the analysis process will be implemented by powerful tools. It promotes the use of IoT-based data by BDA and its processes. In order for the IoT-based device to work better, Rathore suggested a four-layer model [85]. Included are data generation, data processing, and data analysis [85]. It is quoted that IoT data processing should take into account cognitive processing and optimization beyond 2020 [86]. The acquired signal from sensors is obtained by and used in IOT systems for frame-by-frame or batch processing. In addition, the data collected will be used for feature extraction in the IoT method, followed by the classification stage. Algorithms for data classification are used to train computers [87]. Three types of data, supervised, semisupervised, and unattended, can be used for the MAS classification [87].

The cycle of IoT is shown in Figure 2.3. Sensor data are collected. In the filtering process, the data are entered. Cleaning occurs at this level of denotation and data. Therefore, the extraction of features is considered in the classification phase following decisions taken preprocessing based on a detailed learning methodology.

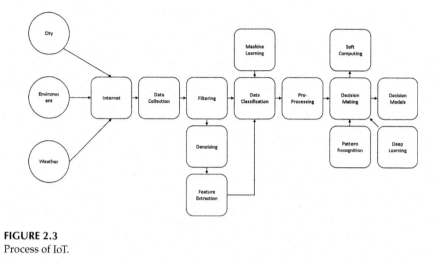

FIGURE 2.3
Process of IoT.

Summary

In this chapter, you will learn machinery and DL from an overview of the big data analysis and its subfields. In order to meet the challenges of data management and give organizations expertise to improve their performance, big data analysis is planned. DNN, RNN, and CNN were presented in this chapter in order to overview deeper learning methods and classification, classifications, and evolutionary techniques. Some techniques from each area have also been investigated. It has also been demonstrated that the application of ML and in-depth learning in IoT-based data makes IoT data analysis more efficient for classification and decision-making.

References

1. Wang, Y., et al., An integrated big data analytics-enabled transformation model: Application to health care. *Information & Management*, 2018. 55(1): pp. 64–79.
2. Nguyen, T., et al., Big data analytics in supply chain management: A state-of-the-art literature review. *Computers & Operations Research*, 2018. 98: pp. 254–264.
3. Jan, B., et al., Deep learning in big data Analytics: A comparative study. *Computers & Electrical Engineering*, 2017. 75: pp. 275–287.
4. Ramsingh, J. and V. Bhuvaneswari, An efficient Map Reduce-Based Hybrid NBC-TFIDF algorithm to mine the public sentiment on diabetes mellitus–A big data approach. *Journal of King Saud University-Computer and Information Sciences*, 2018. doi:10.1016/j.jksuci.2018.06.011.

5. Côrte-Real, N., T. Oliveira, and P. Ruivo, Assessing business value of big data analytics in European firms. *Journal of Business Research*, 2017. 70: pp. 379–390.

6. Esposito, C., et al., A knowledge-based platform for big data analytics based on publish/subscribe services and stream processing. *Knowledge-Based Systems*, 2015. 79: pp. 3–17.

7. Iqbal, R., et al., Big data analytics: Computational intelligence techniques and application areas. *Technological Forecasting and Social Change*, 2018. doi:10.1016/j.techfore.2018.03.024.

8. Gandomi, A. and M. Haider, Beyond the hype: Big data concepts, methods, and analytics. *International Journal of Information Management*, 2015. 35(2): pp. 137–144.

9. Chen, J.-L., The synergistic effects of IT-enabled resources on organizational capabilities and firm performance. *Information & Management*, 2012. 49(3–4): pp. 142–150.

10. Loebbecke, C. and A. Picot, Reflections on societal and business model transformation arising from digitization and big data analytics: A research agenda. *The Journal of Strategic Information Systems*, 2015. 24(3): pp. 149–157.

11. Germann, F., et al., Do retailers benefit from deploying customer analytics? *Journal of Retailing*, 2014. 90(4): pp. 587–593.

12. Kwon, O., N. Lee, and B. Shin, Data quality management, data usage experience and acquisition intention of big data analytics. *International Journal of Information Management*, 2014. 34(3): pp. 387–394.

13. Wamba, S.F., et al., Big data analytics and firm performance: Effects of dynamic capabilities. *Journal of Business Research*, 2017. 70: pp. 356–365.

14. Zhang, Q., et al., A survey on deep learning for big data. *Information Fusion*, 2018. 42: pp. 146–157.

15. Liu, W., et al., A survey of deep neural network architectures and their applications. *Neurocomputing*, 2017. 234: pp. 11–26.

16. Najafabadi, M.M., et al., Deep learning applications and challenges in big data analytics. *Journal of Big Data*, 2015. 2(1): p. 1.

17. Guo, Y., et al., Deep learning for visual understanding: A review. *Neurocomputing*, 2016. 187: pp. 27–48.

18. Yin, Z., et al., A-optimal convolutional neural network. *Neural Computing and Applications*, 2018. 30(7): pp. 2295–2304.

19. Wang, S., et al., Convolutional neural network-based hidden Markov models for rolling element bearing fault identification. *Knowledge-Based Systems*, 2018. 144: pp. 65–76.

20. Acharya, U.R., et al., Automated detection of coronary artery disease using different durations of ECG segments with convolutional neural network. *Knowledge-Based Systems*, 2017. 132: pp. 62–71.

21. Jiang, S., et al., Modified genetic algorithm-based feature selection combined with pre-trained deep neural network for demand forecasting in outpatient department. *Expert Systems with Applications*, 2017. 82: pp. 216–230.

22. Qawaqneh, Z., A.A. Mallouh, and B.D. Barkana, Age and gender classification from speech and face images by jointly fine-tuned deep neural networks. *Expert Systems with Applications*, 2017. 85: pp. 76–86.

23. Shi, X., et al., Tracking topology structure adaptively with deep neural networks. *Neural Computing and Applications*, 2018. 30(11): pp. 3317–3326.

24. Caliskan, A., et al., Performance improvement of deep neural network classifiers by a simple training strategy. *Engineering Applications of Artificial Intelligence*, 2018. 67: pp. 14–23.
25. Zhang, X.-M., et al., An overview of recent developments in Lyapunov–Krasovskii functionals and stability criteria for recurrent neural networks with time-varying delays. *Neurocomputing*, 2018. 313: pp. 392–401.
26. Ruan, X. and Y. Zhang, Blind sequence estimation of MPSK signals using dynamically driven recurrent neural networks. *Neurocomputing*, 2014. 129: pp. 421–427.
27. Jiang, P. and J. Chen, Displacement prediction of landslide based on generalized regression neural networks with K-fold cross-validation. *Neurocomputing*, 2016. 198: pp. 40–47.
28. Miao, Z., Y. Wang, and Y. Yang, Robust tracking control of uncertain dynamic nonholonomic systems using recurrent neural networks. *Neurocomputing*, 2014. 142: pp. 216–227.
29. Gallicchio, C., A. Micheli, and L. Pedrelli, Deep reservoir computing: A critical experimental analysis. *Neurocomputing*, 2017. 268: pp. 87–99.
30. Osipov, V. and M. Osipova, Space–time signal binding in recurrent neural networks with controlled elements. *Neurocomputing*, 2018. 308: pp. 194–204.
31. Peng, H., et al., An unsupervised learning algorithm for membrane computing. *Information Sciences*, 2015. 304: pp. 80–91.
32. Zhou, L., et al., Machine learning on big data: Opportunities and challenges. *Neurocomputing*, 2017. 237: pp. 350–361.
33. Tack, C., Artificial intelligence and machine learning| applications in musculoskeletal physiotherapy. *Musculoskeletal Science and Practice*, 2019. 39: pp. 164–169.
34. Hüllermeier, E., Does machine learning need fuzzy logic? *Fuzzy Sets and Systems*, 2015. 281: pp. 292–299.
35. Gu, X. and P.P. Angelov, Self-organising fuzzy logic classifier. *Information Sciences*, 2018. 447: pp . 36–51.
36. Feng, L., et al., Rough extreme learning machine: A new classification method based on uncertainty measure. *Neurocomputing*, 2019. 325: pp. 269–282.
37. Wu, J.-L., et al., A patent quality analysis and classification system using self-organizing maps with support vector machine. *Applied Soft Computing*, 2016. 41: pp. 305–316.
38. Chou, J.-S. and J.P.P. Thedja, Metaheuristic optimization within machine learning-based classification system for early warnings related to geotechnical problems. *Automation in Construction*, 2016. 68: pp. 65–80.
39. Gonzalez-Abril, L., et al., Handling binary classification problems with a priority class by using support vector machines. *Applied Soft Computing*, 2017. 61: pp. 661–669.
40. Onan, A., A fuzzy-rough nearest neighbor classifier combined with consistency-based subset evaluation and instance selection for automated diagnosis of breast cancer. *Expert Systems with Applications*, 2015. 42(20): pp. 6844–6852.
41. Gou, J., et al., A generalized mean distance-based k-nearest neighbor classifier. *Expert Systems with Applications*, 2019. 115: pp. 356–372.
42. Pan, Z., Y. Wang, and W. Ku, A new k-harmonic nearest neighbor classifier based on the multi-local means. *Expert Systems with Applications*, 2017. 67: pp. 115–125.

43. Peng, Y., W. Kong, and B. Yang, Orthogonal extreme learning machine for image classification. *Neurocomputing*, 2017. 266: pp. 458–464.

44. Ding, S., et al., Extreme learning machine: Algorithm, theory and applications. *Artificial Intelligence Review*, 2015. 44(1): pp. 103–115.

45. Wan, Y., et al., Twin extreme learning machines for pattern classification. *Neurocomputing*, 2017. 260: pp. 235–244.

46. Xu, S. and J. Wang, Dynamic extreme learning machine for data stream classification. *Neurocomputing*, 2017. 238: pp. 433–449.

47. Du, M., S. Ding, and H. Jia, Study on density peaks clustering based on k-nearest neighbors and principal component analysis. *Knowledge-Based Systems*, 2016. 99: pp. 135–145.

48. Lohrmann, C. and P. Luukka, A novel similarity classifier with multiple ideal vectors based on k-means clustering. *Decision Support Systems*, 2018. 111: pp. 27–37.

49. Tang, L., Y. Tian, and P.M. Pardalos, A novel perspective on multiclass classification: Regular simplex support vector machine. *Information Sciences*, 2019. 480: pp. 324–338.

50. Wang, Q., et al., Local kernel alignment based multi-view clustering using extreme learning machine. *Neurocomputing*, 2018. 275: pp. 1099–1111.

51. Huang, J., Z.L. Yu, and Z. Gu, A clustering method based on extreme learning machine. *Neurocomputing*, 2018. 277: pp. 108–119.

52. Yu, S.-S., et al., Two improved k-means algorithms. *Applied Soft Computing*, 2018. 68: pp. 747–755.

53. Caraveo, C., F. Valdez, and O. Castillo, Optimization of fuzzy controller design using a new bee colony algorithm with fuzzy dynamic parameter adaptation. *Applied Soft Computing*, 2016. 43: pp. 131–142.

54. Maulik, U. and S. Bandyopadhyay, Genetic algorithm-based clustering technique. *Pattern Recognition*, 2000. 33(9): pp. 1455–1465.

55. Koonce, D. and S.-C. Tsai, Using data mining to find patterns in genetic algorithm solutions to a job shop schedule. *Computers & Industrial Engineering*, 2000. 38(3): pp. 361–374.

56. Shah, S.C. and A. Kusiak, Data mining and genetic algorithm based gene/SNP selection. *Artificial Intelligence in Medicine*, 2004. 31(3): pp. 183–196.

57. Babu, G.P. and M.N. Murty, A near-optimal initial seed value selection in k-means means algorithm using a genetic algorithm. *Pattern Recognition Letters*, 1993. 14(10): pp. 763–769.

58. Srinivasa, K., K. Venugopal, and L.M. Patnaik, A self-adaptive migration model genetic algorithm for data mining applications. *Information Sciences*, 2007. 177(20): pp. 4295–4313.

59. Sikora, R. and S. Piramuthu, Framework for efficient feature selection in genetic algorithm based data mining. *European Journal of Operational Research*, 2007. 180(2): pp. 723–737.

60. Dorigo, M., Ant colony optimization. In Onwubolu, G.C. and Babu, B.V. (ed.) *New Optimization Techniques in Engineering*. 1991: Springer-Verlag, Berlin. pp. 101–117.

61. Ning, J., et al., A best-path-updating information-guided ant colony optimization algorithm. *Information Sciences*, 2018. 433: pp. 142–162.

62. Tabakhi, S., et al., Gene selection for microarray data classification using a novel ant colony optimization. *Neurocomputing*, 2015. 168: pp. 1024–1036.

63. Mohan, B.C. and R. Baskaran, A survey: Ant colony optimization based recent research and implementation on several engineering domain. *Expert Systems with Applications*, 2012. 39(4): pp. 4618–4627.
64. Kozak, J. and U. Boryczka, Collective data mining in the ant colony decision tree approach. *Information Sciences*, 2016. 372: pp. 126–147.
65. Liu, H., et al., A path planning approach for crowd evacuation in buildings based on improved artificial bee colony algorithm. *Applied Soft Computing*, 2018. 68: pp. 360–376.
66. Mavrovouniotis, M. and S. Yang, Training neural networks with ant colony optimization algorithms for pattern classification. *Soft Computing*, 2015. 19(6): pp. 1511–1522.
67. Ghosh, A., et al., Aggregation pheromone density based data clustering. *Information Sciences*, 2008. 178(13): pp. 2816–2831.
68. Panda, M. and A. Abraham, Hybrid evolutionary algorithms for classification data mining. *Neural Computing and Applications*, 2015. 26(3): pp. 507–523.
69. Hong, T.-P., et al., A multi-level ant-colony mining algorithm for membership functions. *Information Sciences*, 2012. 182(1): pp. 3–14.
70. Zhang, L. and Q. Cao, A novel ant-based clustering algorithm using the kernel method. *Information Sciences*, 2011. 181(20): pp. 4658–4672.
71. Shunmugapriya, P. and S. Kanmani, A hybrid algorithm using ant and bee colony optimization for feature selection and classification (AC-ABC Hybrid). *Swarm and Evolutionary Computation*, 2017. 36: pp. 27–36.
72. Harfouchi, F., et al., Modified multiple search cooperative foraging strategy for improved artificial bee colony optimization with robustness analysis. *Soft Computing*, 2018. 22(19): pp. 6371–6394.
73. Castillo, O. and L. Amador-Angulo, A generalized type-2 fuzzy logic approach for dynamic parameter adaptation in bee colony optimization applied to fuzzy controller design. *Information Sciences*, 2018. 460: pp. 476–496.
74. Taherkhani, M. and R. Safabakhsh, A novel stability-based adaptive inertia weight for particle swarm optimization. *Applied Soft Computing*, 2016. 38: pp. 281–295.
75. Kuo, R.-J., et al., Integration of growing self-organizing map and bee colony optimization algorithm for part clustering. *Computers & Industrial Engineering*, 2018. 120: pp. 251–265.
76. Bonyadi, M.R. and Z. Michalewicz, *Particle Swarm Optimization for Single Objective Continuous Space Problems: A Review.* 2017: MIT Press, Cambridge.
77. Delice, Y., et al., A modified particle swarm optimization algorithm to mixed-model two-sided assembly line balancing. *Journal of Intelligent Manufacturing*, 2017. 28(1): pp. 23–36.
78. Verma, O.P., D. Aggarwal, and T. Patodi, Opposition and dimensional based modified firefly algorithm. *Expert Systems with Applications*, 2016. 44: pp. 168–176.
79. Wang, H., et al., Firefly algorithm with neighborhood attraction. *Information Sciences*, 2017. 382: pp. 374–387.
80. Wang, H., et al., Randomly attracted firefly algorithm with neighborhood search and dynamic parameter adjustment mechanism. *Soft Computing*, 2017. 21(18): pp. 5325–5339.
81. Kiziloz, H.E. and T. Dokeroglu, A robust and cooperative parallel tabu search algorithm for the maximum vertex weight clique problem. *Computers & Industrial Engineering*, 2018. 118: pp. 54–66.

82. Martí, R., et al., Tabu search for the dynamic bipartite drawing problem. *Computers & Operations Research*, 2018. 91: p. 1–12.

83. Bożejko, W., et al., Parallel tabu search for the cyclic job shop scheduling problem. *Computers & Industrial Engineering*, 2017. 113: pp. 512–524.

84. Silva, M.R. and C.B. Cunha, A tabu search heuristic for the uncapacitated single allocation p-hub maximal covering problem. *European Journal of Operational Research*, 2017. 262(3): pp. 954–965.

85. Rathore, M.M., et al., Urban planning and building smart cities based on the internet of things using big data analytics. *Computer Networks*, 2016. 101: pp. 63–80.

86. Lee, I. and K. Lee, The internet of things (IoT): Applications, investments, and challenges for enterprises. *Business Horizons*, 2015. 58(4): pp. 431–440.

87. Shanthamallu, U.S., et al. A brief survey of machine learning methods and their sensor and IoT applications. In *2017 8th International Conference on Information, Intelligence, Systems & Applications (IISA)*. Golden Bay Beach Hotel, Larnaca, Cyprus, 28–30 Aug 2017. IEEE.

3

Data Collection, Storage, and Retrieval

Paulo Martins and Ali Soofastaei

Types of Data

Understanding how to differentiate the various types of data is an essential first step to perform a data analytics cycle. The approach used to extract, handle, and analyze data is strictly linked with the type of data in hand. Moreover, the format in which the data are presented has a crucial role in facilitating the comprehension of the problem and the observed phenomenon being measured. People who start the analytics process dedicating enough time to understand the type of data they are working with may save precious time by avoiding the selection of inefficient models [1].

In the context of Big Data analytics (BDA), the most general classification of data types comprises two groups:

- **Structured data** – This is the type of data that can be organized in a tabular format – columns and rows. Its elements can be grouped containing related information, which facilitates store, searches, and analysis activities. The most common way to store the structured data is a relational database, usually managed by a relational database management system ("RDBMS"). A Structured Query Language (SQL) is used to access and manipulate the data contained in the relational database. Structured databases were, for many decades, the only easily accessible source of business analytics. Typical examples include financial transactions and data collected from machine sensors. Nowadays, structured data are estimated to represent less than 20% of the available data [1–3].

- **Unstructured data** – Unlike structured data, this type of data is not presented in an organized and hierarchical format – predefined data models or schemas. Usually, it can be stored in NoSQL (not only SQL) databases, data lakes, and data warehouses. Due to the absence of structure, it is complex to manage, search, and interpret, which is the reason why companies have not explored it until the most recent technological advances in Big Data (BD) and artificial intelligence.

Unstructured data are everywhere, and it represents more than 80% of the currently available data in the world. Common examples of sources of unstructured data include social media (text, video, audio), e-mails, server logs, and mobile data [1–3].

It is also common to refer to the semistructured data as an intermediary classification that encompasses some characteristics of both structured and unstructured data. Usually, it presents an unstructured format with some organizational properties and classification characteristics. For example, a twitter post has its unstructured text format but also has an internal structure that allows it to be arranged by date, size, or time.

Sources of Data

A standard classification created by UNECE (United Nations Economic Commission for Europe) defines three types of data sources incorporated in the context of BD: human-sourced, process-mediated, and machine-generated. The next topics summarize each type:

- **Human-sourced information** – This is the type of information source derived from human experiences. In the past, it was maintained in books and works of art, and then transferred to photographs and audio and video files. Nowadays, most of the human-sourced information is available in digital format and stored in computers and virtual environments such as social media platforms and clouds. Many of these data do not follow a management structure [4].

- **Process-mediated sources** – They refer to the type of data originated from business processes-relevant events such as client registration, product manufacturing, a received order, etc. This type of source compromises transactions, reference tables, and relationships, as well as the metadata defining the context. They usually are well structured and stored in relational database systems [4].

- **Machine-generated sources** – The output of every kind of sensors and machines used to quantify and monitor data in the physical world are machine-generated sources. With the impressive growth of sensors and connected devices, high volumes of data are generated, which creates a strategic opportunity for data-driven business. The nature of this source of data is well structured and proper for computational processing as BD technologies are available. Some examples include data from equipment automation sensors, weather, and other environmental sensors, GPS, mobile phone locations, and data from computer systems such as log files and weblogs [4].

Critical Performance Parameters

Before the BD era, the traditional data analytics activities did not face a significant limitation to deliver the expected outcomes. The analytics process was based on the data warehouse structure, Extract, Transform, and Load processes, and a variety of software tools [5]. After the latest development of technologies related to the internet of things, BD, and Cloud Computing, it is possible to collect all necessary structured and unstructured data generated in all phases of an organization's production chain [6]. The task of processing vast volumes of data from multiple sources very quickly demands fast transfer rates and stable and reliable methods. This challenge is even higher in projects designed for real-time purposes. The traditional relational databases have considerable performance problems to process a large amount of unstructured data, such as e-mail, documents, videos, or audio. The best alternative to address these challenges has been NoSQL databases coupled with some performance-monitoring metrics.

Before defining the most important properties for a database specification, it is essential to mention some aspects about the most common measurement unit, time. The definition of time in the analytics of BD perspective evolves a period required to perform a specific or set of analytical activities. A popular nomenclature is "cycle time," which can be defined as the time spent to process all activities from the activation to the end [7], transforming inputs into outputs. Essential elements of a cycle time are "waiting time" – process idle time (from activation until starting, including time spent for querying) and "processing time" (from the starting until completion) [5,7].

A professional responsible for the development and monitoring of systems created under the structure of a NoSQL database must consider these three critical parameters:

Consistency – A database is considered consistent when the entire reading and writing tasks are performed automatically (without breaks and following a predetermined sequence), avoiding partial/incomplete updates to the database. If an error occurs in any step in the process, none of the previous steps is updated, and the entire process needs to be executed again. This parameter is defined to guarantee that all users always have the same view of the data [8].

Availability – Like the traditional definition encountered in other fields, database availability refers to the time the system remains operational. In practice, every time a user requests a valid read or write task, an immediate response will be provided regardless of the state of any individual node in the system [8,9].

Partition-tolerance – In the perspective of database operations, partition-tolerance means that the service continues to be executed as programmed even in the case of one or multiple discrete failures. Those failures may include nodes or network disconnection, and message loss on network interconnects. In these situations, the communication between nodes is not prevented.

Usually, BDA applications are built under a distributed computing architecture, which means that a data workload is spread over several nodes in a cluster of servers. This configuration results in a considerable improvement in the processing time, allowing quick insights from a massive amount of data. Hence, it is essential to ensure a network partition tolerance. The CAP (consistency, availability, and partition tolerance) theorem presented by Brewer stated that it is impossible to have both consistency and availability at the same time. Depending on the analytics objective, the analytics system may prioritize consistency and achieve the best possible availability or vice versa [10].

Data Quality

It has been more than fifty years since the expression "garbage in, garbage out" become popular, but we continue to be challenged with data quality issues [11]. To generate value from the analytics efforts, an organization must be able to assess the quality of its dataset technically. When well-structured and high-quality data are available, the entire analytics process can be more efficient and automated. On the other hand, poor-quality data can undermine any sort of data strategy [12]. A recently published Harvard study showed that only 3% of organizations' data reach the minimum quality standards. The research also confirmed that the problem of data quality is far worse than most of the leaders realize. Weak data reflect negatively as bad work and pervade every department, in every industry, at every level, and for every type of information. Unfortunately, the results provided unquestionable evidence that most of the data are terrible [13]. In 2016, a published IBM study estimated that the cost of low-quality data in the United States counted for about $3.1 trillion [4]. Experts report that companies have experienced an impact of 8%–12% of total revenues due to data quality issues [14]. Figure 3.1 presents some of the consequences of data quality on business.

Based on the findings summarized by Marsh [15], Figure 3.2 shows some statistics related to the impact of data quality on projects and organizations.

In a recent survey, the learning company O'Reilly showed that organizations have too many distinct data sources and a lot of inconsistent data, and they do not have the required resources to address data quality issues. Some building blocks, such as metadata creation and management, data provenance, and data lineage, are often not found in the organizations. All survey respondents were asked to select from a list of data quality problems,

Operational costs

Poor data quality increases operational costs
because time and other resources are spent
detecting and correting errors. Not to mention
the impact of embiased decisions

Business users

Negative effects are carried out such as: less customer
satisfaction, increased running costs, inneficient decision-
making processes, lower performance and lowered
employee job satisfation

Organizational culture

Since data are created and used in all daily operations,
data are critical inputs to almost all decisions and implicitly
define common terms in an enterprise, data constitute a
significant contributor to organizational culture

Company trust

Poor data quality also means that it becomes
difficult to build trust in the company data, which
may imply a lack of user acceptance of any
initiatives based on such data

FIGURE 3.1
Business impacts of data quality.

FIGURE 3.2
Data quality impact on organizations, according to Marsh [15].

all those that are applicable to them [15]. Figure 3.3 presents the results in descending order from right to left.

It is noticed that most of the respondents consider the preponderance of data sources as the most challenging problem. Almost 60% of the selected "Too many data sources and inconsistent data" succeeded by "Disorganized data stores and lack of metadata," which was chosen by 50% of respondents. Another significant issue is related to data labeling. "Poorly labeled data" and "Unlabeled data" combined were cited by more than 70% of respondents [15].

From the perspective of BD, the risks associated with low-quality data are even more evident. Many BD properties can directly impact data quality. For instance, Data Variety implies that data from various sources and formats can reside altogether in the same environment. There are conflicts and inconsistent or contradictory patterns among data from different sources. This high diverse condition increases complexity and is more prone to create quality issues. Another relevant property is Data Velocity, as it results in vast

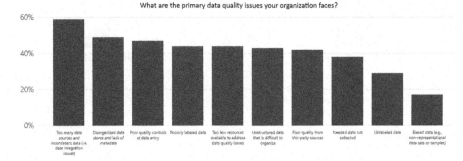

FIGURE 3.3
Primary data quality issues faced by organizations [15].

volumes of data being processed in a short time frame, hence creating more quality parameters to be monitored, such as timeless [16]. It is paramount to measure the impact of data quality to eliminate the risks in the development of BDA solutions.

The Data Quality definition is not only related to its parameters but also involves the business context, including its process and users. For this reason, one of the most accepted concepts is "fitness for use" [17]. Both academia and industry have developed many frameworks and tools to conduct a proper BD quality assessment. In Ref. [18], the authors defend that the concept of BD quality should be adapted for each specific type of BD, considering the different notions of quality.

The next topics will explore data quality assessment and management practices.

Data Quality Assessment

Data quality standards are usually created based on the perspective of data producers, who historically had been the data consumers. However, in the modern world, data users frequently are not data producers, which makes data quality control more difficult. Following this trend, the authors in Ref. [18] used the traditional data quality dimensions and defined their elements and indicators under the view of BD requirements.

The most widely accepted data quality dimensions are as follows:

Availability – determines how affordable and convenient users can collect data and its associated information. It is described by three elements: accessibility, authorization, and timeless.

Usability – refers to the capacity of the data to fulfill users' objectives in terms of definition/documentation, reliability, and metadata.

Reliability –this is the data quality dimension that indicates how reliable the data are and their elements such as accuracy, consistency, completeness, integrity, and audibility.

Relevance – expresses the level of correlation between data content and consumers' desires and requirements.

Presentation quality – refers to a proper representation of the data to make users understand the data.

The methodology was complemented by the assignment of 1–5 elements for each dimension to create a Data Quality assessment framework, as demonstrated in Table 3.1.

Some dimensions and elements can be measured objectively (completeness, timeliness, integrity), and others rely heavily on context or on subjective interpretation (usability, relevance, authorization).

As previously discussed, for each specific business problem and environment, the choice of data quality elements to be used will differ. For each selected dimension, there will be different tools, methods, and processes, and consequently, variations in the required time, costs, and workforce. The first step of the assessment cycle is the definition of the data collection objectives, which are based on business strategy and requirements – for example, operations, decision-making, and planning [18].

Data Quality Strategies

Many data quality challenges are faced in the BD context. It is sometimes difficult to access the data generated from crowdsourcing projects. Machine-generated data often lack enough metadata to evaluate its applicability for a defined process. A range of different approaches for data acquisition, aggregation, and curation is used by different providers without considering the implication for downstream data usage.

Some data quality issues and strategies commonly used to handle them are described in the next lines:

Dealing with Missing Data

Missing data are very critical to BDA. There are plenty of possible approaches, and the simplest one is to eliminate all the missing values from the dataset. However, when a significant portion of the data presents missing values for a critical parameter, its deletion will have an impact on the statistical result. In this situation, an alternative is to use pairwise deletion, which includes the entire dataset for statistical analysis but excludes only the critical parameter. Another standard procedure is to calculate the mean for a group of observations or the entire dataset and replace all missing values with this calculated number. Alternatively, the multiple regression techniques may be applied to predict missing values on a set of highly correlated parameters, although it may cause overfitting in the machine learning models. Finally, a multiple imputation approach may be used combining different methods such as

TABLE 3.1

BD Quality Assessment Framework [18]

Dimensions	Elements	Indicator 01	Indicator 02	Indicator 03
AVAILABILITY	Accessibility	Whether a data access interface is provided	Data can be easily made public or easy to purchase	
	Timeliness	Within a given time, whether the data arrive on time	Whether data are regularly updated	Whether the time interval from data collection and processing to release meets requirements
	Authorization	Whether an individual or organization has the right to use the data		
USABILITY	Definition/documentation	Data specification (data name, definition, ranges of valid values, standard formats, etc.)		
	Credibility	Data come from specialized organizations of a country, field, or industry	Experts or specialists regularly audit and check the correctness of the data content	Data exist in the range of known or acceptable values
	Metadata	Data producers need to provide metadata describing different aspects of the datasets		
RELIABILITY	Accuracy	Data provided are accurate	Data representation (or value) well reflects the true state of the source information	Information (data) representation will not cause ambiguity
	Integrity	Data format is clear and meets the criteria	Data are consistent with structural integrity	Data are consistent with content integrity
	Consistency	During a certain time, data remain consistent and verifiable	Data and the data from other data sources are consistent or verifiable	After data have been processed, their concepts, value domains, and formats still match as before processing
	Completeness	Whether the deficiency of a component will impact data accuracy and integrity	Whether the deficiency of a component will impact use of the data for data with multi-components	
	Auditability	Auditors can fairly evaluate data accuracy and integrity within rational time and manpower limits		
RELEVANCE	Fitness	The data collected do not completely match the theme, but they expound one aspect	Most datasets retrieved are within the retrieval theme users need	Information theme provides matches with users' retrieval theme
PRESENTATION QUALITY	Readability	Data (content, format, etc.) are clear and understandable	It is easy to judge that the data provided meet needs	Data description, classification, and coding content satisfy specification and are easy to understand
	Structure	Refers to the level of difficulty in transforming semi-structured or unstructured data to structured data through technology		

expectation-maximization/maximum likelihood estimation, Markov Chain Monte Carlo simulation, and propensity score estimation. Each method creates a version of the dataset, and those versions are analyzed and combined to produce values estimation and their respective confidence intervals [19].

Dealing with Duplicated Data

It is essential to identify and eliminate duplicated data in any BDA process. Entry errors and data collection methods are some causes of duplication. The identification of duplicated values in big datasets may be complicated. The first issue is because of the assignment of a unique ID to many pieces of information in the same entity (aggregation of all information), named as linking process. The second issue is finding and deleting duplicated data related to the unique IDs [19]. Due to the large volume of data, memory capacity may be limited to process this kind of task. Working with duplicated values can lead to an incorrect interpretation given by the idea that some observations are more frequent than they are. Some methods to deal with this problem involve a visualization step, which provides information about the quantity and pattern of duplicated data, followed by specific functions applied to identify and remove these data.

Dealing with Data Heterogeneity

Variety is one of the five characteristics of BD. The different types of data (structured, semistructured, and unstructured) bring heterogeneity to the context, especially when dealing with Web data and other unstructured and semistructured sources. Some Information Extraction systems have been developed to address these challenges. State-of-the-art in this field comprises sophisticated hybrid models that enhance a combination of statistical and rule-based models. Many references are based on textual data, such as natural language texts. Another trend is focused on extracting text inserted in images and videos. In the future, it is expected to extract information from images and videos (aka feature detection and feature extraction), which depends on the context and is considered highly a highly complex task.

All strategies mentioned above provide a summarized approach to deal with some data quality issues—however, there are many other quality challenges and strategies to handle them. The next section will explore the aspects of Data Quality programs.

Data Quality Programs

According to the DMBOK – Data Management Body of Knowledge [20] – the objectives of Data Quality programs are as follows:

- defining and implementing a data governance approach to guarantee that the generated data are aligned with the data consumer's necessities.
- creating data quality controls based on procedures and specifications.
- establishing a process to measure, monitor, and inform data quality levels.
- identifying improvement opportunities in the process and systems associated with the quality of the data.

Figure 3.4 summarizes the ten fundamental principles that form the basis of Data Quality programs.

Data Quality programs work better when embedded in a Data Governance program. Indeed, data quality challenges usually motivate the necessity of developing an effective Data Governance strategy.

Companies that want to make data quality a priority must start by ensuring that all employees understand the consequences of weak data. A cultural change is required, and the concepts presented in the last sections are essential to succeed in this endeavor.

Data Acquisition

Data acquisition is one of the essential phases of the BDA cycle. As previously discussed, the explosion of data generated from multiple sources in several formats brings complexity and volume that do not fit in the traditional

CRITICALITY
Priorities are defined around the most critical data and its consumers, considering the impact of inaccurate data

LIFECYCLE MANAGEMENT
Management practices are followed across the data lifecycle (from creation to disposal) including internal and external (between systems) movements

PREVENTION
The main goal of DQ programs is avoiding corrective measures by focusing on preventative methods to ensure free errors and high usability of data

ROOT CAUSE REMEDIATION
It is not enough only correcting errors. A deep comprehension of root causes is necessary and often leads to changes in the process and systems

GOVERNANCE
Data Governance activities support high data quality development and DQ program activities must support and sustain a governed data environment

STANDARDS-DRIVEN
Data quality requirements for all stakeholders in the data lifecycle are clearly defined using a measurable approach

OBJECTIVE MEASUREMENT & TRANSPARENCY
Data quality levels need to be measured objectively and consistently and results should be shared with stakeholder, who are the arbiters of quality

EMBEDDED IN BUSINESS PROCESS
The responsibility for the quality of data is assigned to the business process owners, who must ensure data quality standards

SYSTEMATICALLY ENFORCED
System owners must systematically enforce data quality requirements

CONNECTED TO SERVICE LEVELS
Data quality reporting and issues management should be incorporated into Service Level Agreements (SLA)

FIGURE 3.4
Data quality program principles [20].

database management structures. Data acquisition is performed through on-board and off-board sensory and nonsensory data sources. Unlike traditional approaches, BD technologies do not require previous knowledge of the data, meaning before running collection and storage tasks.

The data acquisition process is subdivided into three parts: data collection, data transformation, and data preprocessing. The first step is to collect raw data directly from the source (production environment). Afterward, a high-speed transferring engine is used to send the collected data to the storage system. Finally, data preprocessing techniques are applied to optimize storage capacity by reducing the amount of useless data, such as those redundant data created in environment monitoring systems [21]. In order to reach the acquisition objectives, three main components are illustrated in (Figure 3.5):

Some examples of open protocols commonly applied are the Advanced Message Queuing Protocol (AMPQ), which became the Organization for the Advancement of Structured Information Standards (OASIS) standard in 2012. The AMPQ protocol was developed to reach some industry-specific requirements, such as ubiquity, safety, fidelity, applicability, interoperability, and manageability. Another open protocol is the Java Message Service, which allows a standard way for Java programs to build, transfer, receive, and read an enterprise messaging system's message.

Regarding data acquisition frameworks, Storm is an open-source framework used for distributed real-time processing on streams of data. It can be correctly deployed with different programming languages and storage technologies (relational databases, NoSQL, etc.). Additionally, a standout characteristic of Storm is related to its flexibility to be utilized in different scenarios such as streaming processing and distributed remote procedure call for solving highly intensive computational functions on-the-fly and continuous computation applications. Other examples of frameworks are S4 (Simple Scalable Streaming System), Kafka, Flume, and Hadoop [22].

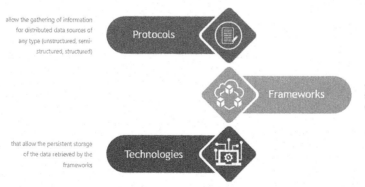

FIGURE 3.5
Data acquisition primary requisites [22].

Although there are many available resources for data acquisition, most of the data are not adequate for processing. The challenge is to increase data quality without extrapolating costs.

Data Storage

The requirements of BD storage systems involve the storage of massive amounts of data, the capacity of supporting high frequency of random writing and reading requests, flexibility, and efficiency to cope with multiple data models. Ideally, storage systems should have unlimited storage capacity and, for security reasons, only accept encrypted data [23].

The process of storing BD involves two essential components: hardware infrastructure and data management. Hardware infrastructure is built of connected information and communications technology resources that are configured flexibly to meet instantaneous demand. A critical requirement for hardware infrastructure is the scalability capacity and dynamic reconfiguration according to the application's characteristics. On the head of the hardware infrastructure is a data management software designed to maintain massive datasets [21].

One of the most common technologies for BD storage is the Hadoop Distributed File System, which allows an analytics interface like that found in Business Intelligence environments. The state-of-the-art in data store technologies is described in Figure 3.6. Although a wide range of technologies exists inside each of these groups, only two examples are provided.

Despite the better scalability of NoSQL databases when compared to relational databases, they may face some limitations when dealing with highly

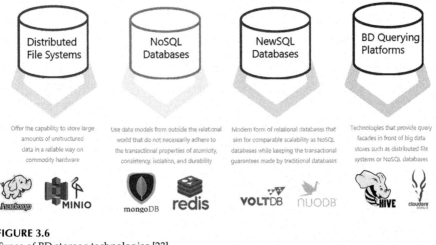

FIGURE 3.6
Types of BD storage technologies [23].

complex data models such as graph databases. One method to solve this issue is to partition the graph into subgraphs. New SQL databases are reported to be approximately fifty times faster than traditional relational systems.

Another fast-growing alternative to BD storage is the use of external cloud storage services. This option has become increasingly popular due to recent decreasing costs, flexible accessibility, high availability, and customizable scalability. Cloud computing enables companies to run essential programs and applications through the internet, reducing time, space, and costs. There are three primary cloud types:

1. **Public cloud** – the infrastructure location is physically separated from the customer. The company that offers the cloud service manages the whole infrastructure and uses shared resources.

2. **Private cloud** – the private type offers the same benefits of Public Cloud but relies on a dedicated, private hardware. The infrastructure is not shared, and the customer has a physical control over the infrastructure. This option ensures the highest security and control levels.

3. **Hybrid cloud** – as the name suggests, the hybrid option is a combination of Public and Private Clouds. Depending on organizational requirements, this integration enables lower costs and performance improvement.

In order to enhance digital capabilities, many companies are considering a cloud-migration strategy on their road map.

Data Retrieval

Data retrieval refers to how the desired data are specified and retrieved from a data store. The process of retrieving relevant information from BD evolves three significant issues: efficiency, effectiveness, and execution time. Data retrieval systems aim to support the storage process and the general organization of the information to meet user's necessities to access the generated documentation. Typically, data retrieval activities begin with a simple query created by the user. In this process, indexing is considered an essential component. The most critical parts of data retrieval systems are as follows: (1) query subsystem, where the user creates a query based on his needs; (2) matching function, applied to compare the query and documents located in the database; and (3) documentary database, which is the place designed to store all the documents [24].

The application of BDA in the information retrieval flow can significantly improve query efficiency allowing easy and effective data retrieval, management of specific databases, and analyzing patterns. The potential lies in integrating traditional data with new forms of data. Data quality issues are also a critical concern in the data retrieval process as decisions depend on obtaining accurate information, and the quality of data is usually highly variable and all too often incorrect. There are many available platforms and a variety of methodologies for data retrieval. The fundamental requirements for platform selection may include availability, continuity, ease of use, scalability, ability to manipulate at different levels of granularity, privacy and security enablement, and quality assurance [25].

Hadoop architecture is commonly used due to its storage capacity, high availability, and retrieval velocity [26]. However, it is considered complex to install, configure, and administrate. Highly skilled professionals are also a critical aspect for many organizations that intend to implement Hadoop [25].

Genetic algorithm (GA) and other evolutionary algorithms have been applied to improve the data retrieval process [24,27]. In Ref. [28], a Deep Learning Modified Neural Networks technique was proposed as a more efficient alternative to existing retrieval methods.

Data in the Mining Industry

The modern mining industry is heavily dependent on several types of data generated continuously in all phases of the value chain. In the exploration phase, drilling campaigns provide the data, which are essential inputs for resource estimation systems that evaluate the economic potential of a mining project. In the operation phase, fleet management system (FMS) is the most abundant source of data used to monitor essential key performance indicators (KPIs), such as productivity and asset performance. To efficiently operate a mineral processing plant, several sensors and measurement systems are implemented to track variables such as equipment vibration and temperature, chemical elements concentration, and particle size distribution. Furthermore, data and its systems are rooted in the supply chain, logistics, marketing, and many other steps of the mining value chain. To extract real-time insights and drive operational excellence, miners are enabled by the declining costs of advanced sensors and the use of sophisticated BD platforms to process this massive amount of data. Justifiably, accurate data are the backbone for the successful digitalization of mining companies.

Standard classification of data sources in the mining industry considers direct and indirect measurements. Direct generated data include those collected from instruments such as Global Positioning System, geodetic surveys, and commodity price monitoring. On the other hand, indirect sources comprise data generated as a byproduct of process and operations, such as drill and blasting, loading, and hauling operations, mineral processing (plant), and logistic systems. Example of those systems includes FMS, supervisory control and data acquisition (SCADA), and distributed control systems (DCSs).

As observed in many other industries, mining faces critical challenges to manage data correctly. It is not surprising for someone to encounter different departments updating multiple copies of the same data or identify corrupted data without knowing why or where. Usually, the major challenge is that there is no single source of truth – central repository, data warehouse, or data lake structure [29]. Moreover, commonly used terms and metrics are often defined using different approaches for each mine site. There is a lack of academic research dedicated to the essential requirements of Big Data management (BDM) in the mining industry [30]. The next sessions will explore the characteristics of data generated in different mining activities and consider quality issues and management practices.

Geological Data

In the geology field, the Geosoft Geoscience Data Management Survey revealed that 60% of the participants declared that geoscientists in their team dedicate more than 20% of the time on data management. Almost 50% of the respondents classified their inability to search and access all geoscience data using one integrated tool as the top challenge [31]. Additionally, a Global Mining Guidelines Group report concluded that geologists spend 10% of their working time converting files. Drilling data is undoubtedly treated as a precious asset for earth resource companies, and high standards and best practices are implemented to guarantee data security and storage effectiveness. Nevertheless, when it comes to extracting value from this data using interpretation and modeling techniques, high standards are not always followed [32].

Data management practices are essential in the portable drill's operation. Future trends involve the integration of sequential activities and data synchronization with the central server to support validation. This multifunctional structure is similar to a mobile survey and sampling laboratory created to extract, access, and analyze a massive amount of manifold and complex geochemical and geophysical data. When Quality Assurance and Quality Control routines are deployed into the data acquisition procedure, data quality issues are identified in advance before leaving the source. The introduction of BDM platforms allows the availability of highly complex and near real-time geochemical/geophysical data for the next phases of the analytics cycle. Finally, the generated results from the analysis are promptly shared near real time to geologists and other stakeholders [30].

Advanced technologies enable the transition from digital geology to intelligent geology by using BD and related technologies to connect various geological systems [33]. In Ref. [34], the authors approached the implications of geological cloud to explore the potential of deep integration and the use of structured and unstructured data from multiple sources such as geology, minerals, geophysics, geochemistry, remote sensing, terrain, topography, vegetation, architecture, hydrology, disasters, and other digitally geological data. Additionally, the entire BDA cycle can be integrated. Some key technologies and trends presented in the article are described below:

Geological BD collection and preprocessing – The objective is to classify all the geological BD collected from geological data, geological information, and geological literature. It can be divided into five sections:

- **Geological data collection access** – It involves the technologies and procedures to quickly access large-scale network information and deliver real time, high concurrency, and fast web content acquisition, according to the cloud requirements.

- **Quality and usability characteristics of geological data** – It is the process of identifying and separating valuable data from irrelevant or low-value data. By applying intelligent discovery and management technologies to filter out these data, storage and processing costs are reduced, and efficiency and accuracy tend to be improved.

- **Geological data entry recognition model** – An entity modeling method is applied to extract entities from the massive quantity of data and discover connections between them. This process ensures that the data are consistent and relevant according to the data model.

- **Aggregation of geological BD** – Aggregation is required due to the heterogeneity characteristic of geological data. It enables data sharing and data fusion between different data sources, allowing data retrieval and data presentation processes unification.

- **Management of geological BD evolution tracking records** – The objective is to track the evolution of BD across the entire life cycle. BD is collected and aggregated to fulfill some critical functions in the Geological Information Service platform.

Geological BD storage management – Handling the considerable variety of geological data in a traditional relational database presents some limitations. Distributed storage technologies such as Hadoop, Spark, and other nonrelational database systems are typically employed to meet specific requirements. Future research topics in the management of geological databases include spatial query considering the benefits of Unified Modeling Language (UML) and Computer-aided Software Engineering (CASE) methodology.

Geological BD analysis and mining – In order to unhide geological knowledge and reach the goal of intelligent mining for geological BD, it is necessary to combine geological data, geological information, and geological literature to understand geological BD environment analysis and mining algorithm. Two aspects are typically considered:

- **Geological BD analysis** – BD technologies are used to create scientifically mathematical models to analyze the metal-organic rules. New geological BD sources extracted from social media and networks are also incorporated to predict future data to drive geological prospecting and providing more effective solutions for geological applications.

- **Geological BD mining** – Data mining techniques are applied to uncover useful insights from the massive multilevel spatiotemporal and attribute data. Multiple approaches include statistics, pattern recognition, artificial intelligence, machine learning, fuzzy logic, advanced visualization, among others.

Highly performable BD cloud computing platform – Distributed computing solutions are the backbone of BDA. Different frameworks present their specific advantages and drawbacks. For example, Hadoop/MapReduce is usually applied for offline complex BD processing, the Spark is used in offline fast BD processing, and the Storm is employed for real-time online BD processing.

Applications of geological BD technologies – BD technologies can interpret the big complex geological data and identify essential patterns that explain the distribution of mineral. In the future, geological data from various adjacent deposits can be connected to identify "digital" characteristics of the distribution of metallogenic materials, access exploration potential, delimit anomalies, and prospective regions.

Operations Data

In the last decade, a variety of sensors and smart chips have been integrated across mining operations. All those sensors and connected devices generate a massive amount of data in real time. The variety of generated data includes geological data, equipment monitoring data, operational performance data, environmental data, among others. An avenue of opportunities has been created with the latest development in Wi-Fi, 4G, and 5G technologies. Drone systems are applied for a variety of activities, such as to measure geotechnical stability, mining mapping, tailings dam monitoring, among others. Real-time insights can be delivered through high-performance computers, artificial intelligence, and machine-learning advanced models.

When working with sensors, standards and procedures must be clearly defined and implemented. Defects or anomalies in instruments can provide inaccurate measurements, produce errors into downstream calculations, impact process control efficiency, and affect user's confidence in information systems [35].

Data provided by FMSs are used to calculate strategic KPIs such as availability, utilization, effective utilization, production, productivity, unproductive time, productive time, mean time between failures, and mean time to repair. In Ref. [36], the author reviewed the consistency of an ordinary FMS extracting a four-month dataset. Haul truck activities' cycle times were assessed, and the samples indicated an inconsistency rate of 21% in the haulage time. Shovel loading times presented data quality issues in 29% of the records. It is essential to mention that the FMS considered is this research is configured to receive manual inputs for most of the assignments, especially cycle time stages (loading, hauling, queuing, dumping, returning, maneuver), maintenance states (corrective and preventive), and unproductive time (shift change, refueling, meals, road maintenance, etc.). This limitation naturally implies a higher tendency for errors when compared with more sophisticated systems developed to detect the majority of trucks and shovels activities automatically.

A similar conclusion was previously reported by Hsu 2015, which assessed datasets from two different FMS vendors in two separate open-pit mine sites. The primary quality problem observed was inconsistent labeling (assignment) of activities to time categories. In terms of quantitative assessment, both systems presented a surprisingly high percentage of concise duration stages, which suggests either data corruption (software/hardware) or human errors (operator input). Additionally, the research identified a mismatch in the maintenance intervention records. All quality issues reported above can potentially undermine any data analytics initiative or continuous improvement project.

A general trend for mining operations data management in the BD era is the integration of different data sources and systems from geological mapping and block model update routines to transportation and disposal of ore (crushing plant and stockpiles) and waste (waste dump). The current dispersed approach loses the value of connecting mine planning software, which contains information about the ore body, FMSs that monitor operational performance and assets' health, enterprise resource planning, which are the general repository of company's information. However, despite all the advantages obtained by enhancing BD capabilities in the mining industry, some topics require further discussion. Data privacy, security, and archiving are significant issues that need serious consideration [37, 38].

Geotechnical Data

Another mining subject that has its specificities is geotechnics. Geotechnical teams are responsible for managing the most critical risks on the mine, such as rock mass failure or collapse and geotechnical structures stability (e.g., tailings dams). In 2017, Anglo Gold Ashanti's experts published a very detailed article to describe data challenges and assessment procedures to evaluate the available geotechnical information in mining sites. The nature of the geotechnical data is not complicated, but geotechnical assumptions rely on the suitability and integrity of data, which creates complexity. The intrinsic unreliability is that the surrounding data affect evaluation and assessment approaches. Accurate data form the foundation to understand rock mass response when a sequence of digging activities is developed. It results in a solid comprehension of the more relevant factors that can drive failure mechanisms, control measures, and strategies to optimize pit design [39]. The following thoughts summarize data management good practices, as suggested in the research.

Data collection standards – A good standard must include all project requirements, equipment limitations, supplier descriptions and responsibilities, and integration strategy considering other systems and protocols. The International Society for Rock Mechanics provides a set of methods and practices to support geotechnical professionals in such a standard.

Securing data – After data collection, there is a need to ensure that it is not lost or modified. Traditional data repositories such as paper records, Excel files, or even access databases are not considered secure. Reasons for this include the ease of losing or change data during edition or transfer tasks – unintended or untracked modifications or the mere deletion of files without a record of the reasons. The best practice is to use a robust Database Management System (DBMS) like the SQL database, which allows configuration for different levels of permissions for access and can log any modification made to the structure and content. A final recommendation is choosing the same database provider used in the geology department to ensure redundancy and workforce skills to execute activities.

Automated data reliability checks – It is essential to map the possible sources of incorrect data and establish procedural routines to check conformance to data integrity. The assessment process is based on the completeness and accuracy of the data records. First, completeness considers whether all data records were collected and inputted correctly. Second, the validation process considers the spatial distribution of the data in the rock mass and its fitness to represent the variety of geotechnical environments. When it comes to accuracy, consistency is essential. This is mainly a challenge due to the degree of interpretation found in some collection methods and the necessity of reflecting reality. For example, data collected from joint rough-ness or orientation properties tend to present a certain level of vari-ability when obtained for different practitioners or even by the same person. In this case, experience and training are crucial to mini-mizing variation effects. Accuracy also involves the elimination of unrepresentative samples and standards for equipment testing and calibration of instruments used for measurements.

Interpretive data reliability checks – An experienced engineer, who understands data collection methods, empirical relations, and char-acteristics of rocks, is the person in charge to perform the interpre-tive validation checks. The outputs of this step are as follows:

- **Validation Report** – Describes errors in the principles of geotech-nical logging data (depth values, missing logging data, wrong logged codes).

- **Stereo plots** – Provides an orientation bias zone for individual boreholes to support the evaluation of missing data.

- **RQD versus spacing** – A RQD spacing chart is applied to com-pare calculated RQD and estimated defect spacing for an interval.

- **Histograms** – Histograms are effectively used for outlier identi-fication and population delineation.

- **Cai GSI graph** – Allows the assessment of data using an interval reference in a graphical framework. It uses the methodology pro-posed by Cai 2004 that compares spacing with the joint condition rating.

- **Q′ factor graph** – Following the same concept of the GSI graph, this graph relates the axes to block size expressed by the ratio of RQD/Jn (y-axis) and Jr/Ja (x-axis), making possible the assess-ment of Q Prime (Q′) which determines the quality of rock mass.

- **Borehole log** – A borehole log is applied to check the distribution of as-logged input parameters to the different rock mass groups, output estimated values, and can be evaluated using lithology or structural context.

Availability of reliable data – It is of utmost importance to ensure that the data are readily available at any given time. ODBC functional-ity (Open Database Connectivity) and spatial functions are used for visualization evaluation of the data.

By using the proposed framework, the authors reported that reliable data were doubled and, in some cases, tripled.

Mineral Processing Data

The primary sources of data in mineral processing plants are collected from Process Control Systems, environmental monitoring systems, and SCADA systems. Advanced sensors and smart devices are present in most of the plant equipment, such as crushers, mills, conveyor belts, flotation cells, and critical components like pumps and pipelines. Most of the data generated are a measurement and a timestamp together, such as energy consumption (kWh) and throughput (t/h).

Mineral processing plants are an exceptional environment for the application of advanced analytics technologies and methods. A wide range of applications is discussed in the literature. Not surprisingly, the quality of the data used for training and testing the models is a critical element. Process mechanics and dynamics are very complex, and it is vastly challenging to take reliable, accurate, or direct measurements of specific process variables [40]. In the perfect world, training data should cover all operating regimes, dynamic plant behavior, have a high resolution, be correctly labeled, and be enough size for machine learning techniques. Mineral processing operations typically produce a massive amount of data. However, often it comes at low resolution (e.g., ore grade measurements, which demand laboratory analysis), with different sampling intervals, containing missing values and noise, and is unlabeled.

According to Ref. [41], a supervisory control platform should be able to perform validation and reconciliation of process data, identify operation and instrumentation issues, and coordinate local control loops under a holistic strategy. Most of these essential capabilities should be defined in the plant design stages, but the major operational parameters usually are only assigned once the process facilities are running. Moreover, a shortage of technical skills is often identified in all managerial and operational levels regarding the comprehension of the modeling and operational aspect of nonlinear and complex processes in dynamic operating conditions [41,42].

The flotation process is considered as one of the most complex and data-sensitive steps in the mineral processing field. Multiple parameters are measured by existing instrumentation such as ore composition, flow rates, and some ore specific properties (e.g., density, pH, pulp levels, and particle size). However, some essential properties such as liberation degree, surface chemistry, bubble size distribution, bubble loading remain challenging to measure and infer [42,43,44].

A set of artificial intelligence techniques have been successfully applied in the creation of smart operational systems, providing a better alternative with a higher tolerance for imprecision, uncertainty, and partial truth. Some frequently used examples are fuzzy logic, artificial neural networks, GAs, support vector machines, decision trees, and hybrids of these methods [45].

By applying some data quality concepts discussed in this chapter, mining organizations can significantly improve their efficiency in all activities from pit to port and support effective, fact-based decision-making.

Summary

Exploring the full potential of BD and its related technologies requires fundamental comprehension of basic data concepts. Before starting any data analytics process, it is essential to understand the multiple data sources available and interdependencies between them. The various types, formats, and magnitude of data are essential factors in the composition of the business context and, consequently, the objective of the analytics program. Additionally, they guide the selection of BD technologies and methodologies to meet the challenges of data acquisition, storage, and processing. A pervasive topic that frequently undermines the success of data analytics programs is Data Quality. Special attention must be dedicated to understand causes and mitigate the effects of weak data. An effective strategy can be created by embedding Data Quality programs into Data Governance programs.

References

1. Ozdemir, S., S. Kakade, and Tibaldeschi, *Principles of Data Science*, 2nd edition. 2018: Packt Publishing, Birmingham.
2. Hurwitz, J., A. Nugent, H. Fern, and M. Kaufman, Unstructured Data in a Big Data Environment. [cited 2020 24/03/2020]; available from: https://www.dummies.com/programming/big-data/engineering/unstructured-data-in-a-big-data-environment/.
3. Marr, B., What's the Difference between Structured, Semi-Structured and Unstructured Data? 2019 [cited 2020 20/03/2020]; available from: https://www.forbes.com/sites/bernardmarr/2019/10/18/whats-the-difference-between-structured-semi-structured-and-unstructured-data/#362053832b4d.
4. Firmani, D., M. Mecella, M. Scannapieco, and C. Batini, *On the Meaningfulness of "Big Data Quality"*. 2015: Springer, Berlin.
5. Mohamed, I., A. Rusli, J. Yusmadi, and R. Nor, A Review on the performance measurement of big data analytics process. 2018: Faculty of Computer Science and Information Technology, University Putra Malaysia.
6. Cheng, Y., C. Ken, S. Hemeng, Z. Yongping, and T. Fei, Data and knowledge mining with big data towards smart production. *Journal of Industrial Information Integration*, 2017. 9: pp. 1–13.
7. Dumas, M., M. La Rosa, J. Mendling, and H. Reijers, *Fundamentals of Business Process Management*, 2nd edition. 2018: Springer, Berlin.
8. Gessert, F., W. Wingerath, S. Friedrich, and N. Ritter. NoSQL database systems: A survey and decision guidance. *Computer Science - Research and Development*, 2016. 32: 353–365.
9. Nazrul, S., CAP Theorem and Distributed Database Management Systems. 2018 [cited 2020 20/03/2020]; available from: https://towardsdatascience.com/cap-theorem-and-distributed-database-management-systems-5c2be977950e

10. Gilbert, S. and N. Lynch, Perspectives on the CAP theorem. *Computer*. 2012. 45(2): pp. 30–36.
11. Redman, T., Data's Credibility Problem. 2013 [cited 2020 24/03/2020]; available from: https://hbr.org/2013/12/datas-credibility-problem.
12. Deloitte. Measuring Data Quality. [cited 2020 24/03/2020]; available from: https://www2.deloitte.com/ie/en/pages/tax/articles/measuring-data-quality.html.
13. Nagle, T., T. Redman, and D. Sammon, Only 3% of Companies' Data Meets Basic Quality Standards. 2017 [cited 2020 24/03/2020]; available from: https://hbr.org/2017/09/only-3-of-companies-data-meets-basic-quality-standards.
14. Redman, T., Bad Data Costs the U.S. $3 Trillion Per Year. 2016 [cited 2020 24/03/2020]; available from: https://hbr.org/2016/09/bad-data-costs-the-u-s-3-trillionper-year.
15. Magoulas, R. and S. Swoyer, The State of Data Quality in 2020. 2020 [cited 2020 24/03/2020]; available from: https://www.oreilly.com/radar/the-state-of-data-quality-in-2020/.
16. Taleb, I., El Kassabi, H. T., Serhani, M. A., Dssouli, R., & Bouhaddioui, C. (2016, July). Big data quality: A quality dimensions evaluation. In *2016 Intl IEEE Conferences on Ubiquitous Intelligence & Computing, Advanced and Trusted Computing, Scalable Computing and Communications, Cloud and Big Data Computing, Internet of People, and Smart World Congress* (UIC/ATC/ScalCom/CBDCom/IoP/SmartWorld) (pp. 759–765). IEEE.
17. Strong, D., Y. Lee, and R. Wang, Data quality in context. *Communications of the ACM*, 1997. 40: p. 103.
18. Cai, L. and Y. Zhu, The challenges of data quality and data quality assessment in the big data era. *Data Science Journal*, 2015. 14(2): pp. 1–10.
19. Gudivada, V. and J. Ding, Data quality considerations for big data and machine learning: Going beyond data cleaning and transformations. *International Journal on Advances in Software*, 2017. 10(1): pp. 1–20.
20. Henderson, D., et al., *DAMA-DMBOK: Data Management Body of Knowledge*, second edition, 2017: Technics Publications, Denville; NJ.
21. Hu, H., Y. Wen, T. Chua, and X. Li, Toward scalable systems for big data analytics: A technology tutorial. *IEEE Access*, 2014. 2: pp. 652–687.
22. Lyko, K., M. Nitzschke, and A. Ngomo, Big data acquisition. In Cavanillas, J., Curry, E., Wahlster, W. (eds.) *New Horizons for a Data-Driven Economy: A Roadmap for Usage and Exploitation of Big Data in Europe*. 2016: Springer, Cham.
23. Strohbach, M., J. Daubert, H. Ravkin, and M. Lischka, Big data storage. In Cavanillas, J., Curry, E., Wahlster, W. (eds.) *New Horizons for a Data-Driven Economy: A Roadmap for Usage and Exploitation of Big Data in Europe*. 2016: Springer, Cham.
24. Irfan, S. and B. Babu, Information retrieval in big data using evolutionary computation: A survey. *International Conference on Computing, Communication and Automation (ICCCA)*, Arizona, USA2016.
25. Malavuru, D., R. Shriman, and S. Vijayan, Big Data Analytics in Information Retrieval: Promise and Potential, 2014.
26. Haneef, I., M. Ehsan, Q. Ghazia, and U. Hafiz, Big data retrieval: Taxonomy, techniques and feature analysis. *International Journal of Computer Science and Network Security*, 2018. 18: pp. 55–59.
27. Abualigah, L. and H. Essam, Apply genetic algorithms to information retrieval using vector space model. *International Journal of Computer Science, Engineering and Applications*, 2015. 5(1): pp. 19–28.

28. Prasanth, T. and M. Gunasekaran, Effective big data retrieval using deep learning modified neural networks. *Mobile Networks and Applications*, 2019. 24: pp. 282–294.
29. Edwards, M., Industry Q&A: Mining Data Management, Analysis and Integrity. 2015 [cited 2020 24/03/2020]; available from: https://www.miningglobal.com/technology/industry-qa-mining-data-management-analysis-and-integrity.
30. Wipro. Application of Big Data Solution to Mining Analytics. [cited 2020 24/03/2020]; available from: https://www.wipro.com/en-BR/natural-resources/application-of-big-data-solution-to-mining-analytics.
31. Geosoft. Geoscience data management report. 2017.
32. Seequent. What Challenges Are Mining Companies Facing in Managing Their Geological Data? 2019. [cited 2020 24/03/2020]; available from: https://www.seequent.com/what-challenges-are-mining-companies-facing-in-managing-their-geological-data.
33. Song, M., Z. Li, B. Zhou, and C. Li, Cloud computing model for big geological data processing. *Applied Mechanics and Materials*, 2014. 475–476: pp. 306–311.
34. Zhu, Y., Y. Tan, X. Luo, and Z. He, Big data management for cloud-enabled geological information services. *Scientific Programming*, 2018. 2018(7): p. 1.
35. Rasekh, A., Hassanzadeh, A., Mulchandani, S., Modi, S., & Banks, M. K. (2016). Smart water networks and cyber security.
36. Martins, P., Aplicação de Cartas de Controle para Análise de Estabilidade de Fluxo Produtivo em um Sistema de Gerenciamento de Frota na Mineração, 2019.
37. Qi, C., Big data management in the mining industry. *International Journal of Minerals, Metallurgy and Materials*, 2020. 27(2): p. 131.
38. Hsu, N., Data quality of fleet management systems in open pit mining: Issues and impacts on key performance indicators for haul trucks fleets. Queen's University, 2015.
39. Seery, J., Data management and geotechnical models. In Eighth International Conference on Deep and High Stress Mining, Perth, Australia from 28-30 March, 2017.
40. Bergh, L., Artificial intelligence in mineral processing plants: An overview. In *International Conference on Artificial Intelligence: Technologies and Applications*, January 24-25, 2016, in Bangkok, Thailand 2016.
41. Bergh, L., J. Yianatos, C. Acuna, H. Perez, and F. Lopez, Supervisory control at salvador flotation columns. *Minerals Engineering*, 1999. 12: pp. 733–744.
42. Houdoin, D., Methods for automatic control, observation, and optimization in mineral processing plants. *Journal of Process Control*, 2011. 21(2): pp. 211–225.
43. Bergh, L. and J. Yianatos, The long way toward multivariate predictive control of flotation processes. *Journal of Process Control*, 2011. 21: pp. 226–234.
44. Oosthuien, D., I. Craig, S.-L. Jämsä-Jounela, and B. Sun, On the current state of flotation modelling for process control. *IFAC-PapersOnLine*, 2017. 50: pp. 19–24.
45. Jovanovic, I., M. Igor, and J. Tomislav, Soft computing-based modelling of flotation processes – A review. *Minerals Engineering*, 2015. 84: pp. 34–63.

4

Making Sense of Data

Amanda Ferraboli, Maycown Douglas de Oliveira Miranda,
and Ali Soofastaei

Introduction

Data Science, Data Mining, and Analytics projects always achieve better results if guided by a process methodology. Whether one opts for the most popular Cross-Industry Standard Process for Data Mining, its refined and extended version Analytics Solutions Unified Method for Data Mining, Knowledge Discovery in Databases, Sample, Explore, Modify, Model, Assess for Data Mining or its method, one thing is sure: no matter what Data Mining methodology you apply, there will always be the need for a Data Preparation step. After an initial iteration to understand the kind of data available and the kind of data needed to address a specific business challenge, the data preparation step is where data scientists construct the bridge between these two. The journey of transforming the available data into a desired and handleable data is not always viable; therefore, previous understanding of the information is essential to confirm the viability of the project.

Data Science applied to the Mining Industry has a vital role in safety improvement, accident prevention, forecasting, maintenance needs prediction, time waste, and cost reduction, boosting productivity, avoiding waste of resources, and tracking regulation compliance. Data Preparation accounts for 50%–80% of a Data Mining cycle, and every detail is essential for a successful implementation of the subsequent phases.

Part I: From Collection to Preparation and Main Sources of Data in the Mining Industry

Taking a closer look at the available data before processing helps to avoid unexpected issues and surprises in the Data Preparation step. So, before giving sequence to this chapter framework, it is essential to

1. assure the mapping and completion of data collection,
2. explore data and ideate data preparation tasks, and
3. complete an assessment of data quality.

Data Preparation is the most extended phase of a Data Mining cycle. It is an essential step to cleaning, integrating, transforming, and reducing the raw data, allowing for better results in subsequent modeling and optimization. The preparation framework, however, depends on earlier steps of Data Understand. The connection between Data Understand and Data Preparation lies in the role that the first exerts over the latter. Understanding the data means shaping the data preparation needs and steps, or in other words, preparing the room for putting hands at work in modifying the data set. Understanding the industry-specific data sources usually accelerates the Data Understanding and Preparation phases, as shown in Figure 4.1.

When it comes to working with data in the mining industry, there is plenty of interesting data sources to look at, much a result of the successful and ongoing adoption of IoT technology in the industry.

According to IBM, there is an expectation that IIOT, the Industrial Internet of Things, will deeply penetrate production units, mills, and plants. The adoption is already taking place through sensors, which are becoming each time smaller, cheaper, and more intelligent. These sensors are capable of generating a significant amount of raw data, which, with proper big data infrastructure, can be analyzed in real time and can generate insights for better decision-making [2].

To have an idea about the potential amount of data generated in the mining industry, Rio Tinto's group executive for Technology and Innovation examined that every haul truck they have is equipped with more than 200 sensors which generate five terabytes of data per day – summing up a total of 4,500 terabytes considering all the fleet. In addition to that, each of their plants is equipped with 20–30,000 sensors. Furthermore, even though data volume is a significant characteristic, more meaningful than that is the information, the knowledge, and insights that can be extracted [3]. There surges the importance of Data Mining.

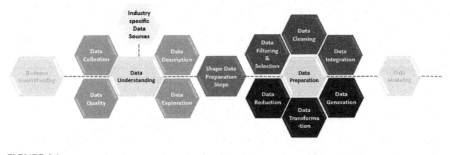

FIGURE 4.1
Data understanding and data preparation detailed connection [1].

As highlighted above, in terms of volume, the Internet of "Big" Things is the source that generates the most significant amount of data. Even though the IoT is a valuable source, it is not the only one. Data can come in different forms and types, and the collection can be performed in a broad spectrum of sources. Figure 4.2 illustrates the primary Data Sources in the universe of the Mining Industry, followed by some practical examples.

Things are corresponding to the group that collects information from vehicles, objects, and locations. It includes industrial and nonindustrial sensors with connectivity capacity: machines, components (e.g., MineCare®), and location and climate (e.g., smart buildings, barometers, seismological activity). Historians are time-series data (e.g., PIMS, Modular Dispatch®, MEMS®) and on-demand reports: maintenance and failure reports (oil quality analysis, tire removal, drilling rigs performance, engines condition analysis, conveyor belts, and railroads inspections).

Plants stand for the data collected in the control and automation process of the mills. Control can also play an essential role at mines in the automated haulage, for example, and is gradually being adopted at ports [4]. However, traditionally, the plants are where this automation layer is mostly active, generating essential process data. Automation data include set points and real-time historical autonomous decisions made by the control and optimization systems.

People include data related directly or indirectly to humans. Human Resources department is often interested in collecting absenteeism, health, and employee performance data to be used in People Analytics reports. Security and Safety departments in the mining industry are very active and robust since the industry is massive and aggressive. Any inattention or mistake may cause severe damages, and discovering patterns in accident and incident data might save lives.

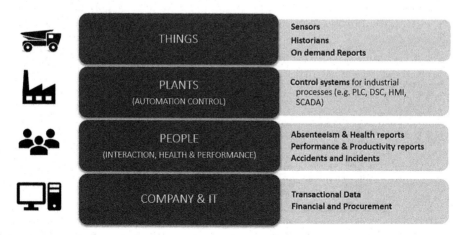

FIGURE 4.2
Main data sources in the universe of the mining industry.

Finally, Company & IT is a source of transactional data. It stores all financial (contracts, invoices, receipts, insurance costs), strategic (plans and records), and logistic data (materials, inventory, storage). It is widespread to have this kind of data stored in SAP applications.

Part II: The Process of Making Data Prepared for Challenges

This section will dive into the Data Preparation steps, techniques, and most common issues. Figure 4.3 illustrates what is meant by data preparation. The main activities in the first part include filtering and cleaning; the second part involves integrating, generating, transforming, and reducing data. The main activities of other phases are also described for contextualization. The presented flow chart in Figure 4.3 provides a big picture of how the Data Preparation phase is incorporated in the data mining cycle, bridging understanding, and modeling processes.

It is further interesting to follow the transformation journey of the information, from initial raw data to cleaned data, then prepared data, evolving to identifiable patterns, and ending as knowledge and insight.

Finally, Figure 4.3 provides a notion of time to consume across the phases. A big part of the effort is dedicated to the initial analysis: Data Understanding and Preparation sum up to 80% of total time. The preparation itself accounts for 50%–70% of the total dedication time.

Many tend to face this phase as a tedious and only time-consuming activity. However, the secret of successful analytics lies much on careful data preparation – from outlier identification and removal, data merging to feature engineering, and dimensionality reduction. A good data scientist is usually curious to mine the data and creativity to propose new data arrangements.

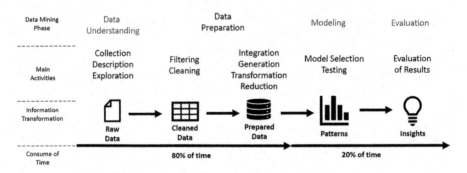

FIGURE 4.3
Data preparation phase in the data mining cycle.

Data Filtering and Selection: Can Tell What is Relevant?

In a data mining initiative, the goals of the project are usually defined at the beginning of the exploratory phase. After understanding the data, one should have a well-built notion about what kind of attributes and records address the defined data mining goals.

As a matter of nomenclature, here, Filtering refers to including or excluding attributes and characteristics while Selecting refers to including or not the records. In other words, columns are filtered, and rows are selected. Take the example of a database containing information on Equipment Physical Availability and Equipment Utilization. Each piece of equipment is a record corresponding to a row. Equipment Type, Date, and the calculated KPIs are all attributes in correspondence to each equipment. Now suppose that the database looks like Table 4.1, and the client is interested in predicting these KPIs for rotary drilling rigs the next month, based on the historical records.

The prediction is oriented toward the rotary drilling rigs only, so a data selection of records must be performed to include only the equipment types which correspond to a rotary drilling rig. By analyzing the tabulated information, it might be concluded that the variable "Cost" has no influence or direct relation with the Physical Availability and Utilization KPIs. To avoid needless processing and visual pollution with insignificant data, it is recommended to filter this column. Table 4.2 highlights the steps described above.

TABLE 4.1

Sample Database

Equipment Name	Equipment Type	Physical Availability	Utilization	Date time	Cost
RDR01	Rotary drilling rig	90	98	2018-01-01	100
SHO01	Shovel	95	80	2017-05-01	200
RDR02	Rotary drilling rig	0,007	0,03	2016-09-17	100
HT01	Haul truck	80	89	2018-05-11	250

TABLE 4.2

Sample Database with Selected Rows and Filtered Columns

Equipment Name	Equipment Type	Physical Availability	Utilization	Date time	Cost
RDR01	Rotary drilling rig	90	98	2018-01-01	100
SHO01	Shovel	95	80	2017-05-01	200
RDR02	Rotary drilling rig	0,007	0,03	2016-09-17	100
HT01	Haul truck	80	89	2018-05-11	250

TABLE 4.3

Final Database After Filtering and Selection

Equipment Name	Equipment Type	Physical Availability	Utilization	Date time
RDR01	Rotary drilling rig	90	98	2018-01-01
RDR02	Rotary drilling rig	0,007	0,03	2016-09-17

The final database should look like Table 4.3.

The decision about the inclusion or exclusion of data must be grounded by precise requirements (e.g., equipment type), hypothesis (lack of connection of variable cost), and constraints. To include a specific attribute, be sure to register the relevance of the quality for the data mining goals and the hypothesis grounded on evidence.

Companies' strategies, systems, metrics, and processes tend to be frequently modified or updated. In terms of constraints, these modifications may turn part of the data invalid for use. Suppose notice a difference in the scale of attributes for the records, for example, the KPIs of equipment RDR01 and RDR02. Continue supposing that the conversion between these scales cannot be done because, until 2016 (date for equipment RDR02), the calculation of Physical Availability and Utilization was based on a variable that does not exist any longer, and as a result, the interpretation of these KPIs changes. This is an incompatibility constraint, and one of the most recommended solutions is to remove all records with Date Time attribute equal to 2016 or earlier. Another common constraint issue involves attributes including sensitive personal information (SPI) and personal information (PI), such as name, address, register number, and credit card information. These attributes are often filtered to comply with ethical principles and data protection regulations.

Data Cleaning: Bad Data to Useful Data

With only the necessary filtered and selected data in hands, now it is time to address inconsistencies and issues that the data might reflect. Bad data are a commonly used term to indicate data that need to be treated. Some examples of bad data are duplicates, missing, zero or particular values, outliers, data errors, measurement error, incomplete, or outdated information.

Duplicates are records registered or included twice or more times in the database. The problem of maintaining duplicates in a database is giving extra weight (bias) to instances with multiple occurrences and generating incorrect results. Common contexts that create redundant data are multiple users feeding the same database or segregated data collection with many separated files.

There are a series of libraries and built-in functions ready to address the duplication issue in Excel, Python, R, or SPSS Modeler to mention some data mining tools. The general idea of these functions is to search for equal values in selected attributes for different records. Attribute selection is customizable, but mostly, the default parameter will distinct records using all attribute columns. The usual recommendation is to consider all columns for identifying duplicates, except record numeric ID, if existent. Table 4.4 highlights data mining tools, most common functions, and methods to remove duplicates.

Another case for duplicates, which is even more challenging, is the duplication of meaning, such as using abbreviations, synonyms, or typos. In these situations, there are advanced approaches, such as using text-mining preprocessing techniques (presented in a later session in this chapter).

Missing values are prevalent to be found in databases. The problem of missing values is that most functions and algorithms fail to execute when a missing value is found. Missing or blank values are often shown as *null* or NA characters, but they can also appear as zeroes or individual values. By the way, zero is a very tricky value because sometimes it represents valid registers, while at other times, it reflects the lack of data. Suppose a database containing mining plant processes data and one of the attributes being the duration of each cycle. On one hand, zero units of duration is an expected value when the process is interrupted; on the other side, if the plant is operating normally and the registration for cycle time is zero, it might reflect an error in sensors or any runtime issue in some system. Particular values might also represent these kinds of errors. For example, in databases, it is not uncommon to find −99, −999, or −999 as a code to represent a missing value. −99 represents the smallest number that could be written using three characters and, for historical, customary (and somehow irresponsible) reasons became a popular, yet unobvious way to represent missing values. To mention another case for special character misleading, excel interprets 0 in date-type attributes as 01/0/1900 or "January 0, 1900."

If the number of missing elements is small, one can remove entire rows or columns with these undesired values. However, removing too many rows/columns might negatively impact the data mining results. Alternative

TABLE 4.4

Most Common Functions and Methods to Remove Duplicates in Data Mining Tools

Data Mining Tool	Duplicates Prevention Method
Microsoft Office Excel	Data > Remove duplicates
Python	Pandas package. duplicated () method (to identify duplicates) Pandas package. *drop _ duplicates ()* method (to drop redundant records)
R	Built-in functions duplicated () (to identify duplicates), unique () (to keep unique records)
IBM SPSS Modeler	Distinct node

correction methods include filling in the missing elements with a randomly selected value from another record or estimated value. Most popular estimated values are mean, median, or mode of the data set (depending on the structure of your data). Nevertheless, the techniques mentioned above are elementary, incapable of representing all nuances of a complex data set.

More robust methods available for filling in missing data are predictive imputation, that is to say, creating a predictive model for the missing attribute based on the available values and predicting the missing data; multiple imputation, which consists of creating multiple acceptable imputations and combining them into a single result; hot-deck imputation, a cauterization method to assign the mean, median, or mode of the cluster to missing values; k-nearest imputation, using the k-nearest algorithm to find the k-nearest neighbors of a record with missing data and assigning the mean, median, or mode of the k-nearest value to the missing field.

Data errors, measurement error, incomplete, or outdated information are issues that can be detected through a careful exploratory data analysis. Understanding the data is crucial in capturing these kinds of errors. For example, if a PIMS database is being prepared for data mining, where one of the attributes indicates the measurement of volume in a tank, one information that should be collected is the nominal capacity of the tank. With that piece of information, it is possible to establish an acceptable range of expected value. Any negative measurements or values that go beyond the range are data errors. Exaggerated or minimum values could represent a unit of measurement inconsistency, and an assessment for conversion should be conducted. In conclusion, for these kinds of problems in data, a logical analysis might be able to find a pattern and allow for the replacement of data, if unviable, the records/attributes with inconsistencies should be removed.

The process of format standardization is a new resource to identify errors or incomplete data (as well as to manage duplicates). Standardize, in other words, is to set a specific content format to attributes. For example, suppose we must register the cause of failure in a particular type of component, let us say a tire. One technician can register a record describing the tire size as "40.00R57," and on the next day, another technician inputs the same tire failure record to the system using the size as "40.00 R57." Any human can understand that with or without the space character, the meaning of the size is the same. However, this additional space character between the last 0 and R is enough to prevent the deduplication from working correctly. A solution for this specific case could be using methods to remove all space characters in the data. However, it is not straightforward, and for those fewer simple cases, standardization helps a lot. For the size of the tire, supposing the terminology is the same for all suppliers and kinds of tires, a standard format could be created of eight characters where the sixth character is necessarily an "R." Any values different from that would be marked as an error.

Outliers are data observations that stand out in the data set because they are very different from most of the data values.

This kind of values often skews the distribution of the data set, provokes asymmetries in the distribution, and might impact statistic measurements like mean or standard deviation, as well as lead to inaccurate interpretations and conclusions about data attributes and unsatisfactory data mining results. It is significant to make a differentiation between the two types of outliers: true values and errors. The generation of outlier errors can come from erroneous data input or registration, measurement errors, or storage/reading corruption of data.

True outliers are extreme yet real values caused by rare phenomena or inadequate sampling. By rare phenomena, an example could be climate conditions.

Nature is something humans can (mostly) predict but rarely control. Suppose that there is an annual data collection where deficient productivity records are found, even lower than the ever-experienced inferior limit of the range. A more in-depth investigation could find out that the values are not wrong accusations but rather a significant and violent typhoon caused all the operations to be interrupted for an entire week, and the production levels were indeed deficient for the period. Inadequate sampling refers to mixing observations from different focuses of studies. For example, if we pick payload data in mining cycles for a specific fleet, the payload should be a value somewhere between the nominal capacity of the equipment (n) and an inferior limit of the range, let us say $(n-100)$. During the data exploration, a case of full payload equal to $n+100$ is identified as an outlier. It might be concluded that the value is not an error nor a simple actual extreme value since the equipment could not resist to this payload. Instead, the findings indicate that the observation corresponds to equipment belonging to another fleet with a nominal capacity equal to $2n$, thus the observation is a valid one, it just does not belong to the specific sampling that we want to analyze.

A widespread mistake is to identify and immediately remove all types of outliers. However, there are essential nuances lost once these records are eliminated without previous analysis. Concerning true outliers, let us imagine the typhoon case and two different data mining situations. The first initiative will use the data to predict the estimated gross production of the next year, as the typhoon is a rare condition and there is no reason to believe the event will happen again; the outlier records should be removed from the data set. The second data mining initiative will make a long-term prediction for maintenance expenditures in the company to assess the viability of a long-term fixed contract. The typhoon occasioned much damage, and a significant amount of money had to be invested in the reconstruction of the site. Even if the typhoon is an infrequent event happening every ten years, as it is a long-term prediction, it might be of interest to the company to include the typhoon-related observations in the estimation. Now regarding erroneous outliers, they might as well be beneficial, specifically, for anomaly detection, for example, in sensors. An extreme sensor value might reflect that this equipment is broken in need of maintenance or that its communication needs to be restored.

The identification of outliers is not a predefined rule as it depends on the data set structure and particularities. However, some statistical methods will help to complete the job. There are plenty of techniques, and some data scientists might have their blended technique for identifying outliers. Two conventional approaches will be presented here, namely, the standard deviation and the interquartile range (IQR) methods. Other techniques to have a look at are Z-score, the robust modified Z-Score, and clustering approaches such as Hautamäki's Outlier Removal Clustering [5].

Standard deviation (SD) is a statistical method to measure data dispersion, or in other words, how the data are stretched. If our data set corresponds to a normal (or Gaussian) distribution, the SD method for outlier identification can be applied. According to Narasimhan, that is because, considering the mean μ, where the peak of the density occurs, and the σ (standard deviation), which specifies the spread or girth of the bell curve, all standard density diagrams satisfy the following properties which is often referred to as the Empirical Rule (a rule based on observations and/or experiments) [6]:

- **68%** of the reflections fall within **1 SD** of the **mean**, i.e., between $\mu - \sigma$ and $\mu + \sigma$.
- **95%** of the reflections fall within **2 SD** of the **mean**, i.e., between $\mu - 2\sigma$ and $\mu + 2\sigma$.
- **99.7%** of the reflections fall within **3 SD** of the **mean**, i.e., between $\mu - 3\sigma$ and $\mu + 3\sigma$.

For a normal distribution, almost all values lie within three standard deviations of the mean, and that is why normally, the observations not belonging to the three standard deviations range will be identified as an outlier. Using three standard deviations is the default outlier cutoff, but this parameter can be customizable. When working with small data sets, the value of two standard deviations could be tested, and on the other hand, when dealing with massive data sets, it might be recommendable to examine four standard deviations.

Figure 4.4 helps to illustrate the concepts. The mean μ for this normal distribution is 10.10, and the standard deviation σ is 3.13. All values between 0.70 and 19.50 are part of the 99.7% of the distribution if the parameters are 3σ. Smaller or larger values that drop outside the three standard deviations range represent the tails' extreme values and would be interpreted as outliers.

Not all distributions, however, are Gaussian-like distributions. For those cases, the IQR is an alternative. Interquartile is also a statistical measure of dispersion, and it corresponds to the difference between the first and the third quartiles, for example, between the twenty-fifth and seventy-fifth percentiles. The IQR is the middle 50% records of the data set. Box plot visualization, for example, is built upon the interquartile measure.

Outlier identification using the IQR is conducted by finding inferior and superior limits corresponding to the values below the first quartile and above the third quartile by the factor k. The general case of IQR

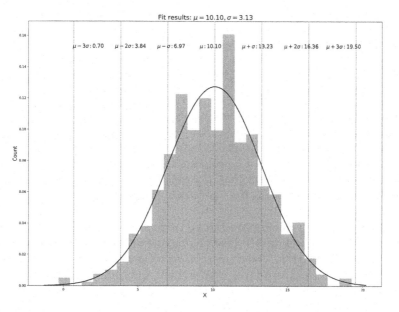

FIGURE 4.4
Normal distribution example.

applied to outlier detection will consider outlier values outside the range: $\left[Q1 - k\,(Q3 - Q1),\, Q3 + k(Q3 - Q1) \right]$, where $Q1$ is the first quartile, $Q3$ is the third quartile, and k is the factor.

A default k value is 1.5, based on the research of John Tukey, known as Tukey's Fences. According to Tukey, any values falling outside the "Inner Fences" (where $k=1.5$) are outliers, and any value falling outside the "Outer Fences" (where $k=3.0$) is an extreme outlier. Tukey's values for k were also based on experiments. The Inner Tukey Fences, for example, is equivalent to 2.7 SDs, with approximately 99.65% of the values falling within the Inner Fences range.

A summary of conditions and proposed solutions for data cleaning is shown below in Table 4.5.

TABLE 4.5

Possible Solutions for Data Cleaning

Data Problem	Possible Solutions
Duplicates	Distinct duplicate records checking all or selected attribute columns; text preprocessing (for duplication of the meaning)
Missing/zero and special values	Remove rows or columns; fill in fields with a random or with an estimated value (imputation techniques)
Measurement errors Incomplete information/ outdated information	Removal of rows or columns; conversion assessment; replacement with a logical or estimated value
Outliers	Careful analysis of the source and impact of the observation; removal using statistical methods

Data Integration: Finding a Key is Key

In most data mining projects, the data sources related to the main question are multiple. That is to say, the final data set that will be used for modeling and analysis is an integration of various kinds of information. For example, studying the influence of fuel composition and consumption and equipment performance would require access to at least productivity, fuel supply, and consumption data. If the data sets contain a common unique identifier in the same granularity, such as an equipment number per hour, it is possible to relate the different data sources. If the data sets do not contain an identifier field, it is necessary to assess the viability of creating this identifier, based on other available attributes.

Two types of data integration can be performed; to mark the differentiation here, distinct names will be designated to each of them: merging and appending. Note that these terms are not a consensus in the literature; there might be other names to refer to one, another, or both.

The main characteristic of a merging process is that records (rows) have a key and similarity, but attributes (columns) are different. The addition is horizontal, in the number of characteristics or qualities. An appending process, on the other hand, has similar attributes (columns) but different records. The addition is vertical, in the number of rows and data points. An example using Tables 4.6–4.8 will visually clarify merging and appending processes. Table 4.6 shows the productivity of one type of equipment per date and shift.

Table 4.7 illustrates equipment names, the weekly amount of fuel supply, and encoded the type of fuel.

Comparing Tables 4.6 and 4.7, it is possible to notice that the records (rows) have a similarity and a key can be identified, even if the name of the column is not the same, the information contained in the column "Equipment ID" from Table 4.6 is the same contained in the column "Equipment Name" from

TABLE 4.6

Productivity of One Type of Equipment per Date and Shift

Equipment ID	Productivity (Daily per Shift)	Operator	Date	Shift
HT05	3,000	4,210	2018-07-01	Night
HT02	4,100	4,315	2018-07-01	Night
HT01	2,650	4,166	2018-07-01	Night

TABLE 4.7

Equipment Names, Weekly Amount of Fuel Supply, and Encoding Type of Fuel

Equipment Name	Fuel Supply (weekly)	Type of Fuel
HT02	530,000	10
HT05	750,000	10
HT01	657,000	10

TABLE 4.8

New Data to Append

Equipment ID	Productivity (Daily per Shift)	Operator	Date	Shift
HT06	4,500	4,457	2018-06-01	Morning
HT04	3,240	4,629	2018-06-01	Morning
HT03	2,950	4,013	2018-06-01	Morning

Table 4.7. The attributes, however, are different. The first table contains data on daily productivity, while the secondary data present a weekly fuel report. Thus, integrating these data sets represents a merging process.

Having a look at Table 4.8, the resemblance to Table 4.6 is immediately recognizable. The records are not the same; equipment numbers are different as well as the dates. However, the attributes match perfectly. Integrating these two data sets means appending rows from Table 4.8 to 4.6.

The expected result from merging is shown in Table 4.9.

Notice that the granularity of productivity and fuel supply is different; one information is collected everyday while the other, weekly. To tackle this challenge, some transformations can be performed, such as aggregation, i.e., summarization of information from multiple records. Imagine the possibility of computing a field that sums up all the daily values to compose the weekly productivity to be comparable with the weekly supply. Aggregation techniques will be presented in a further section. The expected result from appending is shown in Table 4.10.

Table 4.11 highlights data mining tools, the most common functions, and methods to merge different data sets:

TABLE 4.9

Expected Result from Merging Data

Equipment ID	Productivity (Daily per Shift)	Operator	Date	Shift	Fuel Supply (Weekly)	Type of Fuel
HT05	3,000	4,210	2018-07-01	Night	750,000	10
HT02	4,100	4,315	2018-07-01	Night	530,000	10
HT01	2,650	4,166	2018-07-01	Night	657,000	10

TABLE 4.10

Expected Result from Appending Data

Equipment ID	Productivity (Daily per Shift)	Operator	Date	Shift
HT05	3,000	4,210	2018-07-01	Night
HT02	4,100	4,315	2018-07-01	Night
HT01	2,650	4,166	2018-07-01	Night
HT06	4,500	4,457	2018-06-01	Morning
HT04	3,240	4,629	2018-06-01	Morning
HT03	2,950	4,013	2018-06-01	Morning

TABLE 4.11

Data Mining Tools, Most Common Functions, and Methods to Merge Different Data Sets

Data Mining Tool	Merge and Append Data Set Methods
Python	Pandas package. merge () method (to merge two data sets based on a common key) Pandas package. append () method
R	Built-in function merge () (to merge two data sets based on a common key) Built-in functions *c ()* (concatenate) or *append ()* Dplyr package Mutating and Filtered Join set of functions
IBM SPSS Modeler	Merge node and Append node

Data Generation and Feature Engineering: Room for the New

Frequently, there will be the need to go through a Data Generation process, which is a branch of Data Transformation. Generating or constructing new data refers to the process of deriving new attributes or records. The generation of records is less frequent than the generation of attributes; still, it can be beneficial in some cases, like oversampling. Some examples of attribute generation are KPIs and other measures derived from originally collected fields. Some examples of created columns might be speed (distance/time), productivity (amount of work/time), and performance/cost (productivity/ acquisition cost). Sometimes, variables alone are not enough to understand and recommend a solution to a problem, but if they combine two or more variables, a pattern can be observed more clearly.

Features are essential for predictive models because they explain the dependent variable and will significantly reflect the performance of the model. Having these features in good quality and quantity means significantly increasing the chance for good results. However, note that adding irrelevant features to the analysis will not contribute to improving the results.

Feature Engineering is a term frequently used when talking about data generation and transformation for machine learning. In other words, it is the procedure of producing, constructing, and transforming critical attributes.

A few tips about how to go through a meaningful feature engineering process include studying the data and the means to reflect expertise about the topic/industry in the features; combining scarce groups within the records; as mentioned in Filtering and Selection section, removing unused attributes, in order not to add noise or outdated information to the model; including interaction features and adding flag variables.

As an example, consider Table 4.12, where we have productivity, operator, and date. A simple feature of engineering would be to aggregate productivity per operator until the maximum date is observed.

TABLE 4.12

Sample Table for Feature Engineering

Productivity (Daily per Shift)	Operator	Date
3,000	4,210	2018-07-01
4,100	4,210	2018-08-01
2,650	4,210	2018-09-01
4,500	4,457	2018-07-01
3,240	4,457	2018-08-01
2,950	4,457	2018-09-01

TABLE 4.13

Sample Data with Mean, Standard Deviation, Maximum, and Minimum

Productivity (Daily per Shift)	Operator	Date	Arithmetic Mean	Standard Deviation	Maximum	Minimum
3,000	4,210	01/07/2018	3,250.00	617.79	4,100.00	2,650.00
4,100	4,210	01/08/2018	3,250.00	617.79	4,100.00	2,650.00
2,650	4,210	01/09/2018	3,250.00	617.79	4,100.00	2,650.00
4,500	4,457	01/07/2018	3,563.33	672.82	4,500.00	2,950.00
3,240	4,457	01/08/2018	3,563.33	672.82	4,500.00	2,950.00
2,950	4,457	01/09/2018	3,563.33	672.82	4,500.00	2,950.00

Moreover, the result of the aggregation can be used to extract mean, sum, standard deviation, and other statistical measures. Table 4.13 used mean, standard deviation, minimum, and maximum.

Other tables and features could be used to generate new ones, like using information about the work time of each operator and extract the most frequent shift of operator or last n vehicles IDs. The possibilities are very extensive, depending on the business problem, thus the need for creativity, curiosity, time, and guided effort.

Data Transformation

Data transformation is composed of Data Generation (already detailed in the previous session) and other transforming data methods that do not necessarily generate new fields or records; instead, they change the scale of the data. That is why Data Transformation can be divided into Feature Engineering or Data Generation and Feature Scaling. Primary scaling methods include Normalization, Standardization, and others, such as Fourier Transform, Log Transform, Min-Max transform, and Encoders.

Normalization is a way to treat a sample as a vector of features and then divide each feature by the Euclidean norm of the vector so that the vector will have unit norm. Min-Max, also known as the min-max normalization method as it is the most popular normalization technique, refers to rescaling an attribute range to fitting a new range of [0,1] or [–1,1]. It is useful for a couple of reasons, like making the modeling process less sensitive to the range (scale) of attributes or avoiding a considerable variance of the features.

Standardization is often confused with normalization; they might have the same goal to make an entire set of records share the same property, but the implementation is slightly different. This method rescales an attribute such that its $\mu=0$ and $\sigma=1$. It is necessary for various machine learning methods to have optimized results and to compare the features which are on different scales.

Aggregation can also be used as data transformation, and it refers to the summarization of records. Some traditional measures are sum, mean, maximum, minimum, median, variance, and record count. Some types of aggregation can also be performed by making use of unsupervised learning, i.e., clusterization. The utility of aggregation lies in the fact that summarizing data speeds up the training and learning processes and requires less computational power. However, it is important to warn about aggregation cons as well, such as loss of distribution information.

Encoders are a set of tools to map information, such as encode labels into numbers, and it is very important as many machine learning algorithms work only with numeric data.

An example of data transformation is shown in Table 14.14, where the data set has a column with nonnumeric data and a correspondent column with this information encoded into numbers.

Table 4.15 shows some feature scaling and encoder techniques that can help to achieve better results in a data mining project [7].

Data Reduction: Dimensionality Reduction

Dimensionality in machine learning and data-related fields refers to a data set number of attributes, or in other words, how many columns are being considered as explainable variables and added to the modeling step. It is essential to find the best low-dimensional representation of the data and avoid making models unnecessarily complicated, prevent overfitting, and

TABLE 4.14

Sample Data with Nonnumeric Data and Its Encoded Form

Equipment ID	Productivity (Daily per Shift)	Operator	Date	Shift	Shift Encoded
HT05	3,000	4,210	01/07/2018	Night	1
HT02	4,100	4,315	01/07/2018	Night	1
HT01	2,650	4,166	01/07/2018	Night	1
HT01	4,000	4,200	02/07/2018	Day	0

TABLE 4.15

Common Feature Scaling, Encoder Techniques, and Recommended Use

Feature Scaling Technique	Recommended Use
Normalization	The scaling vector to its unit form is useful whenever a distance-based algorithm is used. Also, when the data set has sparse features (many zeros), this scaling method can be useful.
Standardization	Useful to deal with a multivariate analysis that has different scales. Also, artificial neural networks better converge when the data are in the standard scale form.
Fourier transform	Useful to deal with high-dimensional time-series data.
Log transform	Useful when the data set has skewed features.
Min-Max	Useful to achieve better convergence of optimization algorithms. Gradient descent algorithm is highly sensitive; scaling the data to a specific range can help to speed up.
Encoders	Can be used to map a set of values to another, especially the case of string data; an encoder can be used to map string values to numeric values

the inclusion of redundant variables. There are two main approaches for optimizing dimensionality: feature selection and feature extraction.

Feature extraction refers to simplifying multidimensional data, maintaining fewer selected dimensions, and at the same time, a meaningful representation. The final reduced representation should have a dimensionality that corresponds to the essential dimensionality of the data, which is the minimum number of parameters needed to maintain the observed properties of the original data set. The most popular methods are principal component analysis, single-value decomposition, and Kernel PCA.

Feature selection methods find a smaller subset of a many-dimensional data set to create a data model. The primary strategies for feature selection filter techniques (using mainly statistical measures like Pearson correlation), wrapper techniques (using a predictive model), and embedded techniques (which perform feature selection while building a model, mainly represented by regularization methods). Since feature extraction techniques change the original feature representation of the data and, consequently, result in more limited interpretability, feature selection tends to be an interesting alternative.

Part III: Further Considerations on Making Sense of Data

Unfocused Analytics (A Big Data Analysis) vs. Focused Analytics (Beginning with a Hypothesis)

How can the data be described in terms of Volume, Variety, Velocity, and Veracity? An in-depth exploration can be done through big data when it is still not clear what the hypothesis is that wanted to test or if there is no explicit knowledge about the data and therefore does not know where to start from or what kind of data could be discarded. Working with big and complex data sets is now possible and viable due to advances in computing. From a large amount of data, it is possible to uncover hidden patterns, extract correlations, and deliver many other insights to orient the conduction of the studies.

Usually, programming languages and traditional data mining software are not prepared for dealing and processing big, huge, and gigantic data sets. Also, transferring and extracting these huge data sets can be very difficult and time-consuming. In the mining industry, historians and other types of data are collected in a very high frequency, generating terabytes of data. Some tools and frameworks worth mentioning for this kind of issue are Apache Spark and Hadoop. Both are idealized to deal with elastic scalability, and all the data exploration and preparation steps can be conducted directly in the data storage. Usually, there is a Data Engineer, responsible for the architecture and environment, who will work together with data scientists and data analysts to develop a big data project. Data engineers tend to focus on software engineering, database design, production code, and ensure that data are generally flowing from the source, from where they are collected, to the destination, where they are processed by data scientists.

However, when there is an initial hypothesis to attack the problem or if the partner business team can help to provide some of the hypothesis they want to validate, the work will have a focused analysis. Knowing the question, having the answer is the best strategy for effective analytics. It is essential to collect the questions and the answers that are wanted to be found. The quality of the questions will be crucial to the quality of the insight that takes as a result.

Many data transformations can occur to promote the tests of the established hypothesis. The critical point is to investigate and to verify if a statement about the data can be checked with the available information. Moreover, if not, how can one work on to reach this goal by applying techniques to augment the knowledge about the data, like using a different scale for a numeric field, or generate new variables using the current ones.

The big step in this phase is to maintain a goal in mind and analyze which kind of data is needed to validate some hypotheses; this can be done in many ways, for example, looking for a correlation between numerical data.

Time and Date Data Types Treatment

Time and date information are two standard terms in the real world, and they are important to operations because of factors such as sequence, planning, scheduling, ordering, and elements that contribute both to analyze historical behaviors as well as identify patterns that can be predicted.

Another name to date and time variables is *"Date Time"*; this type is often stored as strings, and therefore, the following information about it may not be present, and it could not be possible to retrieve sorted date by example. Many high-level programming languages offer proper types to store time and data, which do preserve the temporal elements and have functions to do some mathematical and another kind of operations with date and time.

Computer systems also have a configuration of date and time format that vary accordingly to time zones. This can impact the analysis, leading to problems and errors, and cause some mistakes in the interpretation of the information. Consider as an example a system that uses the following data format: "MM-DD-YYYY," like "02–09–1992" (February 9th, 1992), if no information about the data format was provided, this could be wrongly interpreted as "DD-MM-YYYY," and inserted as "02–09–1992" (September 2nd, 1992), while the correct date would be "09-02-1992" (February 9th, 1992).

Ensuring that the format of Date Time variable is correct is an important step, and dealing with Date Time fields requires transformations to enable the use of these variables with its temporal elements. A set of steps can be used to identify the process needed to obtain data that fulfill the data mining goals, also avoiding some known problems. These steps are related to technical and nontechnical situations, like applying a function to properly format the data or gathering data from different time zones, respectively. Consider Table 4.16 as an example of a data set with Date Time variables. Using a Date Time variable turns more practical to execute aggregation in the data by grouping a fixed time interval, like a group of one day.

Results after the aggregation are shown in Table 4.17; note the average value of each sensor by each equipment throughout the day:

Date Time variables also have a significant influence when working with time series analysis. Whenever a time series problem comes up, it is associated with periodicity or time, and it is possible to work with the temporal

TABLE 4.16

Data set with Date time Variables

Equipment Name	Equipment Type	Sensor	Value	Date time
HT01	Haul truck	A	0.51	2018-01-01 04:05:34
HT02	Haul truck	A	4.17	2018-01-01 04:05:34
HT01	Haul truck	A	8.85	2018-01-01 06:05:34
HT02	Haul truck	A	7.34	2018-01-01 06:05:34
HT02	Haul truck	B	4.03	2018-01-01 06:05:34
HT02	Haul truck	A	6.77	2018-01-02 00:00:00
HT01	Haul truck	B	1.28	2018-01-02 00:00:00
HT01	Haul truck	B	2.26	2018-01-02 04:05:34
HT02	Haul truck	A	7.20	2018-01-02 06:05:34
HT01	Haul truck	B	6.91	2018-01-02 16:00:00
HT01	Haul truck	A	3.52	2018-01-02 16:05:34

TABLE 4.17

Results After Aggregation

Equipment Name	Sensor	Date time Date	Average Value
HT01	A	2018-01-01	4.68
HT01	A	2018-01-02	3.52
HT01	B	2018-01-02	3.48
HT02	A	2018-01-01	5.78
HT02	B	2018-01-01	4.03
HT02	A	2018-01-02	6.99

TABLE 4.18

Data Represented as a Time Series

Equipment Name	Value	Date time
HT01	10	2017/01/01
HT01	10	2017/01/02
HT01	5	2017/01/05
HT01	4	2017/01/06
HT01	9	2017/01/07
HT01	7	2017/01/08
HT01	6	2017/01/09
HT01	7	2017/01/10
HT01	6	2017/01/11
HT01	10	2

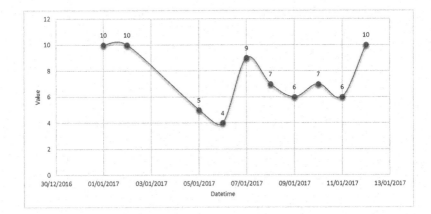

FIGURE 4.5

Time series plot of values for equipment.

distribution to model the data to represent a time series, like in Table 4.18, where there is a sensor value for each day for each equipment.

There is the possibility to plot this .data and to visualize how the observation values vary throughout time, like in Figure 4.5

TABLE 4.19

Transposed Data

Equipment Name	Window	Date time Value 1	Date time Value 2	Date time Value 3	Date time Value 4	Date time Value 5
HT01	1	10	10	5	4	9
HT01	2	7	6	7	4	10

It is also common to transform the data into windows, transposing the temporal information to create different analyses, as shown in Table 4.19. Thus, the various points in time are used as inputs, i.e., explanatory variables, for the model.

Dealing with Unstructured Data: Image and Text Approaches

Not all kinds of data are gathered in a structured form, that is, information with a defined data model and labels. Sometimes it is required to generate a model to the data before using the information. This can be treated as a preprocessing step called feature extraction, where some data transformation is done.

Unstructured data can be described as information without a data model, and two of the most frequently found examples of it are text and images.

Text information, when in its raw forms, such as body email, free description of machine behavior, free-type maintenance/service form, and other types of text data, is unstructured text. It is important to note that text information, in some cases, can also be structured. For example, a text form that requires only a selection of a value from a list of options. In this case, there is a data model, mapping the field in a domain. Figure 4.6 illustrates

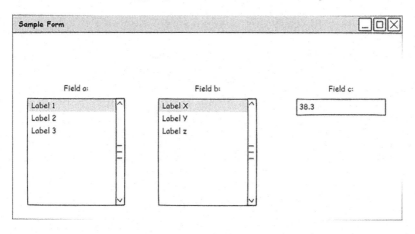

FIGURE 4.6
Sample form to illustrate the structured text.

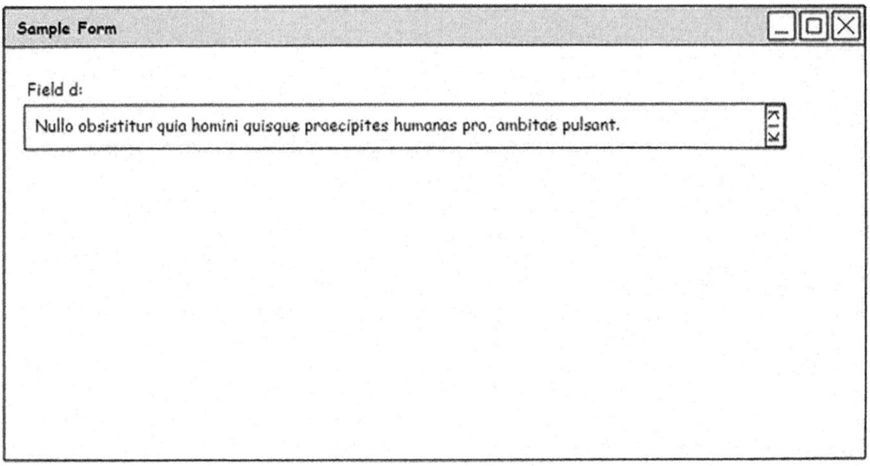

FIGURE 4.7
Sample form to illustrate the unstructured text.

an example, where a form shows three fields, a, b, and c, and two of them, a and b, are of text type. However, there is a list of possible options from which to select the answer. In this case, each field can receive a single label from a list.

In other cases, text can be treated as a type of information that does not contain any data model linked to it, and consequently, no map between field and domain. Like in Figure 4.7, where a free-type text form is displayed so that the user can input text on it, regardless of a list with predefined values of labels. Each of these answers is called a *document*, and a set of documents is called a *corpus*.

To perform an analysis using unstructured text, one must conduct a series of preprocessing steps to obtain a proper data frame. Several methods can be used to reach this goal: an interesting and simple one is the *bag-of-words*.

In summary, *bag-of-words* is a data model used to extract information from the corpus (text) and transform it into a vector representation. This transformation is performed by counting the occurrence of each word in the text and creating a matrix to store the result. This matrix very often receives the name of the *term-document matrix*.

Considering three documents as follows:

$$doc1 = \text{"Data mining is cool!"}$$

$$doc2 = \text{"text mining is coolier."}$$

$$doc3 = \text{"data mining for mining is awesome."}$$

To proceed with the generation of a term-document matrix, one can simply create a dictionary from all documents, using each unique word as the key, and the number of occurrences of the word in the document as the value.

$$docl = \{\text{"awesome"}: 0, \text{"cool"}: 1, \text{"coolier"}: 0, \text{"data"}: 0, \text{"Data"}: 1,$$
$$\text{"for"}: 0, \text{"is"}: 1, \text{"mining"}: 1, \text{"text"}: 0$$

$$doc2 = \{\text{"awesome"}: 0, \text{"cool"}: 0, \text{"coolier"}: 1, \text{"data"}: 0, \text{"Data"}: 0,$$
$$\text{"for"}: 0, \text{"is"}: 1, \text{"mining"}: 1, \text{"text"}: 1$$

$$doc3 = \{\text{"awesome"}: 1, \text{"cool"}: 0, \text{"coolier"}: 0, \text{"data"}: 1, \text{"Data"}: 0,$$
$$\text{"for"}: 1, \text{"is"}: 1, \text{"mining"}: 2, \text{"text"}: 0$$

The following step consists of translating the dictionary into a vector form:

$$docl = [0, 1, 0, 0, 1, 0, 1, 1, 0]$$

$$doc2 = [0, 0, 1, 0, 0, 0, 1, 1, 1]$$

$$doc3 = [1, 0, 0, 1, 0, 1, 1, 2, 0]$$

And, finally, transforming these vectors into a matrix:

$$TD = \big[[0, 1, 0, 0, 1, 0, 1, 1, 0], [0, 0, 1, 0, 0, 0, 1, 1, 1],$$
$$[1, 0, 0, 1, 0, 1, 1, 2, 0]\big]$$

Now, the term-document matrix can be used as an input to an advanced technique, such as a classifier.

Other preprocessing steps can be conducted to improve the results of text-mining tasks. In the previous example, words like "Data" and "data" have the same meaning, but one is written with a capital letter. Data transformations in the text before the generation of the term-document matrix can include lowercasing all the words, removing punctuation, removing numbers from the text and stripping accents, stemming, or lemmatization.

Another kind of feature extraction in a raw text is the annotations. An example of an annotation is the part-of-speech tagging (POS tagging, POST).

This technique extracts a word from the text and classifies it grammatically, according to the context of the speech. This can be useful to reduce ambiguity and also to contribute to context information about the data. To illustrate this case, consider the example below:

doc1 = "Data mining is cool!"

doc1 = {"Data":"noun,""mining":"noun,""is":"verb,""cool":"adjective"}

It is important to notice that these are just a few of the most popular resources when it comes to data preparation for text mining, hundreds of other techniques exist and can be combined, and the application will depend on the data mining goals to be achieved [8].

The second type of unstructured data addressed in this chapter is the images. Although an image may present a structure and have meaning, such as in Figure 4.8, this meaning is not stored within the image and therefore is not part of the data.

We can see Figure 4.8, retrieve some information about its context, and decide to recognize a drawing of a haul truck. However, a machine grounds image recognition mainly on the structure of the image. That is why it is so common that a machine frequently confuses and misinterprets images. A classical and hilarious case involves machines being fooled with pictures from Chihuahuas and muffins [9]. What is evident to a human is not always so visible to a machine.

Some common extensions of images like .jpg and .png are treated as matrices of numbers. A .jpg image with size $M \times N$ and RGB channel color is a matrix of dimensions $M \times N \times 3$, where M denotes its height, N its width for each channel, an image of .png extension has one more channel, *alfa*, and a .jpg image in grayscale has only one channel. Each element $a_{i,j,k}$ of the matrix is a *pixel*, and it is the most granular information of an image [10].

FIGURE 4.8
Sample image, a type of unstructured data.

TABLE 4.20

Common Methods to Deal with Unstructured Data

Data Mining Tool	Unstructured Data Type	Recommended Open-source Methods
Python	Images	Sk image. feature module OpenCV (Open source computer vision package)
R	Images	Magic package
Python	Text	Sk learns. Feature _ extraction module
R	Text	TM package

The value of pixel represents the intensity of the respective channel in a point of the image, and depending on the analysis, each image can be an element of data set, or each pixel of the image, as well as each region can also be an element composing a data set.

When dealing with image data, the pixel information is used as input, and some preprocessing might be necessary before using the data, such as applying digital filters and operators to the images. Using filters and operators is a way to extract information from images such as edges of the elements, level of shadow, brightness, and contrast. In Table 4.20, there are some of the recommended methods to deal with unstructured data.

Summary

In this chapter, we started analyzing methodologies for conducting Analytics and Data Mining projects and concluded that Data Preparation is a single step in all frameworks. In sequence, Data Preparation was contextualized in the Data Mining cycle, bridging Data Understanding, and Modeling phases. The importance of data analysis in the mining industry was grounded by real-life examples of how much data the industry generates in a short period. Decision-makers in the mining industry need to have visibility of all possible data sources they can collect data from. This chapter highlighted some of the most common sources and types of data. From machines to people, the mines of today are highly connected, and a massive opportunity for performance improvement arises from this advent.

Data Democratization is a tendency for the next decades, consisting of enabling nonspecialists to collect and analyze data without the need for help. The phenomenon considers both the viability of access to data as well as the development of a culture of using data. Data-related activities once focused on IT teams are becoming increasingly decentralized to HR, Finance, Supply, Security, and other areas. A Data Democratization project aims to provide business areas with affordable, autonomous, and efficient decision-making, increasing the potential of delivery in all areas.

Data analysis might seem a very technical activity at first glance, but with little guidance, every analyst and decision-maker can become a "data literate" and start mining data sets. Precisely, this chapter intends to play this role of guidance. However, it is important to say that the chapter does not explore all the possibilities of data preparation; instead, the main goal is to generate an initial interest in exploring data; for those readers who are interested in excavating their information and knowledge from a universe of material, articles, and references that are yet to be explored in this (data) mining journey.

References

1. Shearer, C., The CRISP-DM model: The new blueprint for data mining. *Journal of Data Warehousing*, 2000. 5(4): pp. 13–22.
2. Merry, H., 5 benefits IoT is having on the mining industry. 2017.
3. Lilleyman, G., *Innovation in Mining – A Template for Australian Industry*. 2016. RioTinto: Brisbane. p. 10.
4. Chu, F., S. Gailus, L. Liu, and L. Ni. The Future of Port Automation. 2018 [cited 2019 30/05/2019]; available from: https://www.mckinsey.com/industries/travel-transport-and-logistics/our-insights/the-future-of-automated-ports.
5. Hautamäki, V., et al., Improving k-means by outlier removal. In *The Scandinavian Conference on Image Analysis*. Joensuu, June 19–22, 2005. Springer.
6. Narasimhan, B., Normal Density. 1996 [cited 2019 30/05/2019]; available from: http://statweb.stanford.edu/~naras/jsm/NormalDensity/NormalDensity.html.
7. Mörchen, F., Time series feature extraction for data mining using DWT and DFT. 2003, Univ.
8. Manning, C., Understanding human language: Can NLP and deep learning help? In *Proceedings of the 39th International ACM SIGIR Conference on Research and Development in Information Retrieval*. 2016. ACM.
9. Togootogtokh, E. and A. Amartuvshin, Deep learning approach for very similar objects recognition application on Chihuahua and Muffin problem. arXiv preprint arXiv:1801.09573, 2018.
10. Hollink, L., et al., A corpus of images and text in online news. In *LREC*. 2016.

5

Analytics Toolsets

Russell Molaei and Ali Soofastaei

Statistical Approaches

Statistical approaches are generally analysis methods where the general aim is to find a small number of data properties, such as average and variance of data, that represent different properties of the data. The selection of a statistical model is not an easy and straightforward choice. The thought of all data sets as providing a suitable methodology is incorrect. Could the modeling tool answer a particular question? The choice of statistical method can be calculated by the relations between the dependent parameters and the explanations. It can be useful to analyze these interactions graphically. Such models can often be spherical, polynomial, or nonlinear and can be more fitting than linear models. The selection of a statistical model can also be strictly linked to the precise business issue. The problem then, in many cases, is how many variables should be introduced into the process after selecting the correct modeling method. A more significant number of variables is perhaps an excellent method for adapting the model to the data. A methodology that disproportionately provides the data often shows the method used and less precise interpretations of the whole population. Technical quality analysis is the use of some standard indices, including the Bayesian Information Criterion (BIC, SBC) or Akaike's Information Criterion (AIC), as a stabilization between fair data enforcement and a small numbers of variables. The technique with the lowest index has the best value in the collection when associating all parametric techniques together.

Statistical Approaches Selection

The selection of an appropriate approach uses a statistical model and is the most important phase in an analytics project. Table 5.1 can be useful in choosing a practical and accurate statistical model in different situations.

In the following, a couple of useful statistical approaches will be introduced. These methods are widely used in mining literature in connection with data analysis.

TABLE 5.1

Statistical Approaches [1–5]

Dependent Variable	Explanatory Variable(s)	Parametric Models	Conditions of Validity	Other Solutions
One quantitative variable	One qualitative variable (= factor) with two levels	One-way ANOVA with two levels	1; 2; 3; 4	Mann–Whitney test
	One qualitative variable with k levels	One-way ANOVA	1; 2; 3; 4	Kruskal–Wallis test
	Several qualitative variables with several levels	Multi-way ANOVA (factorial designs)	1; 2; 3; 4	
	One quantitative variable	Simple linear regression; nonlinear models (depends on the shape of the relationship between the dependent/explanatory variable)	1–3	Nonparametric regression (*); quantile regression; classification/regression trees (*); K Nearest Neighbors (*); classification/regression Random Forest(*)
	Several quantitative variables	Multiple linear regression; nonlinear models	1–6	PLS regression (*); K Nearest Neighbors (*)
	The mixture of qualitative/quantitative variables	ANCOVA	1–6	PLS regression (*); quantile regression; classification/regression trees (*); K Nearest Neighbors (*); classification/regression Random Forest (*)
Several quantitative variables	Qualitative and/or quantitative variable(s)	MANOVA	1; 4; 7; 8	Redundancy analysis; PLS regression (*)

(Continued)

TABLE 5.1 (*Continued*)

Statistical Approaches [1–5]

Dependent Variable	Explanatory Variable(s)	Parametric Models	Conditions of Validity	Other Solutions
One qualitative variable	Qualitative and/or quantitative variable(s)	Logistic regression (binomial or ordinal or multinomial)	5; 6	PLS-DA (*); Discriminant Analysis (*); classification/regression trees (*); K Nearest Neighbors (*); classification Random Forest (*)
One count variable (with many zero's)	Qualitative &/or quantitative variable(s)	Log-linear regression (Poisson)	5; 6	

(*) solutions designed more for prediction

Conditions of validity

1. Individuals are independent.
2. Variance is homogeneous.
3. Residuals follow a normal distribution.
4. At least 20 individuals (recommended).
5. Absence of multicollinearity (if the purpose is to estimate model parameters).
6. No more explanatory variables than individuals.
7. Multivariate normality of residuals.
8. Variance is homogeneous within every dependent variable. Correlations across dependent variables are homogeneous.

Analysis of Variance

Analysis of Variance (ANOVA) is a parametric statistical technique used to compare the datasets. This technique was invented by R.A. Fisher, hence it is also referred as Fisher's ANOVA. It is used to test discrepancies in a sample between two or more classes. ANOVA offers a statistical test in its purest form to decide whether population means are equal in multiple groups. In contrast, all groups consist of a random sample of a single population. ANOVA is appropriate for comparing (testing) the statistical significance between three or more groups [3]. The most common technique used for comparison of means is, by far, this technique. To employ ANOVA, the first sum of squares or between-groups variance is calculated as [4]

$$SS_{\text{between}} = \Sigma n_j \left(\overline{X}_j - \overline{X} \right)^2 \tag{5.1}$$

where $\overline{X}_j$ is jth group mean, $\overline{X}$ denotes overall mean, and n_j is the sample size per group. Then, from the groups, the degree of freedom, is calculated using

$$df_{\text{between}} = m - 1 \tag{5.2}$$

Moreover, using equations (5.1) and (5.2), the mean of the square between them is:

$$MS_{\text{between}} = \frac{SS_{\text{between}}}{df_{\text{between}}} \tag{5.3}$$

Now, within-group variances are calculated as follows:

$$SS_{\text{within}} = \Sigma \left(X_i - \overline{X}_j \right)^2 \tag{5.4}$$

where $\overline{X}_j$ represents a group mean and X_i denotes an individual observation. For n independent observations and m groups,

$$df_{\text{within}} = n - m \tag{5.5}$$

Then the mean square within is calculated as

$$MS_{\text{within}} = \frac{SS_{\text{within}}}{df_{\text{within}}} \tag{5.6}$$

The F-test is eventually used to compare the total difference factor

$$F = \frac{MS_{\text{between}}}{MS_{\text{within}}} \tag{5.7}$$

The expected value of F is 1 for no difference between the means of the groups. As the value of F increases above 1, the evidence is increasingly inconsistent with the null hypothesis.

ANOVA was utilized in different kinds of studies to address mining engineering problems. For example, for the assessment of mine roof falls [6], evaluation of the volume of dust exposures during mining activities [7], and optimizing the compressive strength of regulated, low-filter materials [8].

The ANOVA expansion will proceed as a consequence of other useful statistical methods, such as analysis of covariance (ANCOVA) and multivariate variance analysis (MANOVA) [4].

Study of the Correlation

Correlation is a bivariate technique used to analyze the strength of the relationship between two continuous variables, calculated numerically, and the direction of the relationship. This analysis is useful when a researcher wants to establish if there are possible connections between variables [9]. The variables are stated to be connected when the movement of another variable accompanies the change of one variable.

Commonly, correlation analysis is utilized to determine if any change in an independent variable is due to a variation in a dependent variable. So, it provides a measure of the correlation between one or some dependent variables and one or some independent variables [10].

Regarding the strength of the connection, the importance of the relationship coefficient fluctuates between +1 and –1. A value of ± 1 shows a perfect degree of dependence among the two variables. As the correlation coefficient value tends 0, the relationship between the two variables will be weaker. The coefficient indicator shows the direction of the association. The correlation between two variables may be positive or negative.

Generally, in statistics, four types of correlations are measured: Pearson correlation [11], Kendall rank correlation [12, 13], Spearman correlation [14–16], and the Point-Biserial correlation [17, 18]. The Pearson Product Moment correlation, the most widely used correlation coefficient, is introduced in the following [19, 20]. The correlation coefficient between X and Y variables is calculated by the formula

$$\text{corr}(X,Y) = \frac{\text{cov}(X,Y)}{\sigma_X \sigma_Y} \tag{5.8}$$

where σ_X and σ_Y are standard deviations of X and Y variables and cov(X, Y) is covariance of X and Y variables, calculated through

$$\text{cov}(X,Y) = \frac{\Sigma\left[(X-\bar{X})(Y-\bar{Y})\right]}{n-1} \tag{5.9}$$

It is important to note that there may be a nonlinear relationship between two continuous variables, but calculating a correlation coefficient does not detect this. Therefore, before computing a correlation coefficient, it is always necessary to carefully evaluate the data. Particularly useful are graphical displays to explore connections between variables.

Correlation Matrix

The correlation matrix demonstrates the relationship among sets of parameters [21, 22]. The correlation matrix of n random parameters $X_1,..., X_n$ is the $n \times n$ matrix whose (i, j) entry is corr(X, Y). Each random variable (X_i) in the matrix is correlated with each of the other values in the table (X_j). This helps to see which pairs of variables have the highest correlation (Table 5.2).

The matrix of correlation is symmetrical because the relationship between X_i and X_j is the same as the correlation between X_j and X_i. Moreover, the diagonal of the table is always a set of ones, because the relationship between a variable and itself is always 1.

Three wide explications are available on how to process a matrix of correlation.

- Patterns on a wide range of data.
- For further comments to be included. For example, the relationship between a target variable and several independent variables is analyzed in order to select a subset of separate variables as the input for further analysis and as a diagnostic for advanced analysis.
- For example, with linear regression, high correlations indicate that linear regression results are inaccurate as a diagnostic when checking specific tests.

Reliability and Survival (Weibull) Analysis

Weibull analysis (also called *life data analysis*) aims to predict the life of a product. By fitting a statistical distribution to live (failure) data from a representative sample of the product population, Weibull analysis makes predictions

TABLE 5.2

Correlation Matrix of Five Variables (Sample)

	X_1	X_2	X_3	X_4	X_5
X_1	$X_{1,1}$	$X_{1,2}$	$X_{1,3}$	$X_{1,4}$	$X_{1,5}$
X_2	$X_{2,1}$	$X_{2,2}$	$X_{2,3}$	$X_{2,4}$	$X_{2,5}$
X_3	$X_{3,1}$	$X_{3,2}$	$X_{3,3}$	$X_{3,4}$	$X_{3,5}$
X_4	$X_{4,1}$	$X_{4,2}$	$X_{4,3}$	$X_{4,4}$	$X_{4,5}$
X_5	$X_{5,1}$	$X_{5,2}$	$X_{5,3}$	$X_{5,4}$	$X_{5,5}$

about life (failure) expectancy of the product [23]. In order to estimate the vital life characteristics of a product, for example, likelihood of or certain failure in a particular time, mean life, and failure rate, the parameterized distribution for the data set is used. To perform a Weibull analysis, the following steps should be considered [24]:

- collecting life (failure) data of the product;
- choose the distribution period that corresponds to the model and the product life data;
- predict the variables which match the data distribution; and
- develop tests and plots that estimate the product's life characteristics, including reliability or mean life.

The Weibull system can be used in a variety of ways, including a parameter, two parameters, three parameters, or a Weibull mixture. The lognormal, exponential, and regular distributions are also commonly utilized in practice. The analyst selects the life distribution most appropriate for modeling each data set based on health and experience.

The 3-parameter Weibull pdf is given by

$$f(t) = \frac{\beta}{\eta} \left(\frac{t-\gamma}{\eta} \right)^{\beta-1} e^{-\left(\frac{t-\gamma}{\eta} \right)^{\beta}} \tag{5.10}$$

The Weibull is a versatile distribution that takes on the characteristics of other forms of distributions, based on the amount of the form parameter, β.

where $f(t) \geq 0, t \geq \gamma, \beta > 0, \eta > 0, -\infty < \gamma < +\infty$. Moreover, η is a scale parameter or characteristic life, β denotes shape parameter or slope, and γ is location parameter or failure-free life.

Figure 5.1 shows the effect of different values of the shape parameter, β, on the shape of the Weibull distribution pdf. The model takes on a variety of forms based on the amount of β, as is seen in Figure 5.1.

In order to adapt a statistical model to a live dataset, the analyst approximates the life-long variables that best fit the results. The variables govern the pdf function's form, scale, and position. For example, in the three-parameter Weibull model (shown above), the scale parameter, η, defines where the bulk of the distribution lies. The shape parameter, β, determines the form of the distribution, and the location variable, γ, describes the position of the timely distribution.

Variables that match a lifetime distribution to a data set are calculated by many different methods. Several available variables evaluation techniques include maximum likelihood estimation probability plotting, rank regression on x, and rank regression on y. Depending on the data set and, in some cases, the chosen life cycle may lead to alteration of the correct analysis technique.

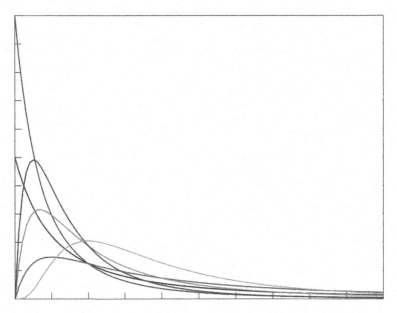

FIGURE 5.1
Weibull pdf with different amounts of β [23].

The results of lifetime analysis can be provided in several forms, including:

- **Mean life** – The median time before the failure of the units in the population.
- **Probability plot** – A image of the probability of time failure.
- **Reliability given time (RGT)** – The probability of a machine operating successfully at a given time. For instance, after three years of operation, the product is 65% likely to work successfully.
- **Failure rate** – The number of failures per unit time can be calculated for the product.
- **Probability of failure given time** – The choice of a device stopping at a certain point in time. The capacity for failure is also known as reliability, and reciprocity is reciprocal. For example, after three years of operation, there is a 35% risk that the unit will fail and a 65% chance of operation (reliability).

Evaluations based on the tracked lives of several units are life-analysis consequences; the tests are unreliable due to the limited sample dimensions. "To calculate this uncertainty because of a sample error," also called "confidence intervals," is used to express the trust that the number of interests is in a given interval. It is not known whether or not a given interval contains the number of interests.

Two-sided or unilateral are the two ways of expressing trust limits. Two-sided links are added to suggest that a particular trust is kept in the amount of interest. The one-sided limit is used to demonstrate that the amount of interest above the lower bound is higher. Besides, with exact confidence, a lower bond is seen. The correct bound type depends on the entry. For example, the analyst should apply unilateral lower reliability and unilateral higher reliability for percentage failure below guarantee and a bidirectional boundary of distribution variables. For example, the 95% smaller two-sided band is the 95% lowest single bond, and the 95% most massive double-sided band is the 95% larger unilateral bond.

This method was used in many different studies on mining engineering to conduct a reliability analysis of mining equipment and for rock failure analysis [25].

Multivariate Analysis

Multivariate analysis (MA) is used to analyze more complex data sets than can be treated by standard analytical methods. It is based on the statistical theory of multivariate statistics, including the measurement and analysis simultaneously of two or more statistical results variables. The MA combines the effects of all parameters on the output and performs the analysis on multiple dimensions [26,27].

By combining all variables in MA, a whole image will be produced regarding the relationship between variables and outputs.

There are many ways to perform an MA. The main objectives of such analysis are [27]:

- Dimensionality reduction;
- Density clustering; and
- Nonlinear distribution modeling.

Some of the previously used multivariate analysis methods in mining engineering are as follows:

- Principal Component Analysis;
- Factor analysis;
- Canonical Correlation;
- Equality of Covariance;
- Discriminant Analysis;
- MANCOVA;
- Correspondence Analysis; and
- Log-linear Models.

State-Space Approach

The state-space approach is established on explaining a time-varying procedure by a vector of quantities. These quantities are collectively called the state of the process. The evolution of the procedure over time is signified as a trajectory in the space of states, for example, a successive transition from one state to another.

The state-space model is a representation of the dynamics of a system that is of Nth order. The first-order differential equation is called the state. Generally, a continuous-time linear dynamic system which is multi-input/multi-output can be expressed by

$$\dot{x}(t) = A(t)x(t) + B(t)u(t)$$
$$y(t) = C(t)x(t) + D(t)u(t)$$

(5.11)

where t is time parameter, $x(t)$ is the vector of state, $u(t)$ is the input or control variable, and $y(t)$ is the output of the system or dependent variable. Also, $A(t)$ is the dynamic matrix, $B(t)$ is the input matrix, $C(t)$ is the output of the sensor matrix, and $D(t)$ is the feedthrough matrix.

In most cases, the systems are inherently linear time-invariant. Hence, the A, B, C,, and D matrixes of (5.11) are not time-variant. So, they are constant.

$x(t)$ is known as the state of the system at time t. This means that the future output of the system is connected only to current state and future inputs to the system. Also, it depends on past input only through the current state of the system. The state is like a memory that summarizes the result of previous inputs on future output.

The common uses of state-space approaches are as follows:

- Modeling and control;
- Forecasting; and
- Clustering and classification.

State-Space Modeling

There are some approaches to create a state-space model of a system. The easiest way is to from Nth-order differential equations that describe the dynamics, according to Equation (5.11). There are a lot of different choices of state parameters, and the only issue is which set of states to use. Also, the state-space model can be developed from a transfer function model.

Linear state-space models can be developed from nonlinear system dynamics. It is a crucial feature of state-space models as most of the systems have nonlinear dynamics. A nonlinear set of dynamics can be given by

$$\dot{x} = f(x, u) \tag{5.12}$$

where $f(x, u)$ is a nonlinear vector function that expresses the dynamics of the system.

The first step is to define the point about which the system is somehow linear, to perform linearization. This occurs around equilibrium points.

The equilibrium point is a point for which if the system starts there, it will remain there for all future time. To calculate the location of the equilibrium point, it is required to set the state derivative (5.12) to zero. The result is an algebraic set of equations that needs to be solved for both x_e and u_e, where e denotes the equilibrium point. By definition, $\dot{x}_e = 0$ and $\dot{u}_e = 0$. These equilibrium points act as set points or operating points for the nonlinear system, and it is about these points that the system dynamics is linear.

State-Space Forecasting

State-space models can be deployed for forecasting purposes. This is a very general approach that can involve differing forecasting algorithms, including regression and ARIMA. It can also include a Bayesian approach to forecasting and predictive models with time-varying coefficients [28].

State-space models are primarily based on the Markov property, which expresses that the future of a process is independent of its past, having the present system state. In this type of system, the state of the process at the current time involves all of the previous information that is necessary to predict the future output of the system [29].

Assume that the state vector expresses the system state at time t x_t. The elements of this vector are denoting the state of the system at time t and do not necessarily observe a state-space model comprised of two equations. First, an observation or measurement equation that formulates how time-series observations are produced from the state vector. Second, a state or system equation that formulates how the state vector evolves through time. These two equations can be written as

$$y_t = h'_t x_t + \varepsilon_t \tag{5.13}$$
$$x_t = Ax_{t-1} + Ga_t$$

where h_t is a known vector of constants, ε_t is the observation error, A and G are known matrices and, a_t is the process noise. If the system is multivariate, then y_t and ε_t will be vectors and the vector h_t changes to a matrix H.

The state-space formulation is not a forecasting technique. In other words, it deploys other forecasting methods; some are described in Equation (5.4). The state space approach utilizes a Bayesian formulation of the problem in which the model parameters have a prior distribution. Using the observation data, this prior distribution is updated into a posterior distribution.

Another formulation helps the coefficients in the regression model to vary through time.

The state-space approaches provide a common mathematical framework to be used for modeling and forecasting. It also allows the relatively easy generalization of many distinct models, some of which are described in this chapter.

Predictive Models

Predictive models (PM) is the term used for the procedure of developing probability and data mining to predict future outcomes. A predictive model utilizes several independent parameters that are predictors that are likely to influence the desired dependent variable (forecasting output). When data has been gathered for appropriate predictors, a statistical algorithm is expressed. The model can employ a simple linear equation, or it may be a sophisticated machine learning algorithm like a neural network. Predictive modeling mainly overlaps with the field of machine learning, and many of the algorithms that are utilized to develop forecasting models form part of the machine learning and artificial intelligence contexts. PM is regularly associated with weather forecasting, online advertising, and marketing. However, it has many uses in mining engineering.

One of the most frequently overlooked challenges of predictive modeling is obtaining the right data to apply when developing algorithms. Data collection and preparation is the most challenging step to develop a predictive model. It is essential to find the best predictors to feed into the model. A descriptive analysis of the data and data treatment, including missing values and outlier fixing, is a very crucial task that takes most of the time needed for predictive modeling.

It is essential to know that big data does not make predictive models more accurate. According to the mathematical theorems, after a certain point, feeding more data into a predictive analytics model does not improve accuracy. Sampling representative portions of the available data may help to develop the model more quickly.

After collecting data, the selection of the right model is important for the next steps. Linear regression includes the most accessible types of PMs. However, there are many complex artificial intelligence algorithms. The complexity of the model does not guarantee the performance of the prediction. Model selection should be considered in connection with

- Data availability and quality; and
- Forecasting period.

After modeling, estimation of performance should be done to measure the accuracy of the model.

Some of the well-known predictive modeling methods, widely used in mining engineering literature, are presented in this section.

Regression

Regression is a type of analytical modeling technique that investigates the relationship between a dependent (target) and independent variable(s) (predictor) [30]. This technique is used for predicting, time series modeling, and finding the causal effect relationship between the variables. Regression Analysis (RA) is an essential method for modeling and analyzing data [31]. In this method, a curve is fitted on the data points, where the differences between the distances of data points from the curve are minimized. RA will be explained in more detail in the following sections.

There are multiple advantages to using RA, such as the following:

- RA demonstrates significant relations between dependent and independent variables;
- RA measures the intensity of the impact on a dependent parameter of several independents;
- RA helps researchers in various scales to compare the effects of calculated parameters;

Such benefits help market analysts to eliminate and determine the best set of variables for the construction of PMs (Figure 5.2).

A variety of tools are available for predicting regression. The three main methods (number of separate variables, type of variables based, and regression line form) drive these methods. There are seven regression types in total: Linear, Logistic, Polynomial, Stepwise, Ridge, Lasso, and Elastic Regression.

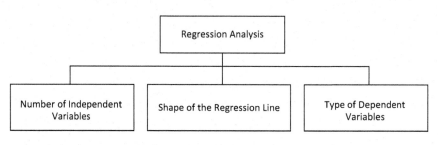

FIGURE 5.2
Regression modeling.

Linear Regression

Linear regression (LR) is the simplest method utilized for prediction. LR forecasts a dependent variable's expected value (Y), as the linear combination of one or more independent variables given unknown quantity (X) and a set of observed values (predictors) and establishes a relationship using a best-fit straight line (also known as a regression line). Figure 5.3 demonstrates a regression line that models the data (blue points).

LR models aim to find a linear mapping between predicted values and real dependent values so the model can predict the dependent variable by having new instances of the independent variable.

A simple LR is used the case for an independent variable. If more than one variable is present, the process is called multiple linear regression.

The following equation represents this:

$$Y = Xa + \text{err} \qquad (5.14)$$

where Y is a vector of dependent variables, Y_i $(i = 1,..., n)$, X is the matrix of independent variables, a matrix of row-vectors x_i or n-dimensional column-vectors X_j, a is the $(n+1)$-dimensional parameter vector, and *err* is a vector value of error. This element of the model is named disturbance term, the error term, or sometimes noise. This equation will be applied to predict the value of the target variable based on a given predictor's parameter(s).

The best fit for the observation data is determined by reducing the sum of the squares of the vertical variations of each data point to the curve, according to the Least Square method, the best method used for fitting a return line.

$$\min_{\text{err}} Xa - Y^2 \qquad (5.15)$$

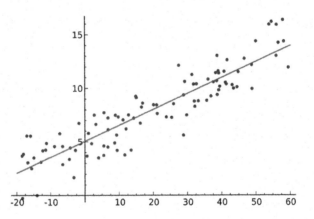

FIGURE 5.3
Linear regression.

So, we need to aim to reduce the sum of the errors (err), due to the best fitting line. To use a linear regression model, some critical assumptions should be considered, as follows:

- A linear relation between independent and dependent variables. A linear configuration of the parameters (regression coefficient) and the predictor variables is the mean of the response parameter. However, predictor variables can be added, and there can be a transformation of available predictors and predictor variables. This trick is applied, for instance, in the Polynomial Regression Method (PRM) that applies LR to match the response parameter as an arbitrary PRM function of a predictor parameter. This creates LR, an extremely powerful inference method.

- Multiple regression suffers from multicollinearity, autocorrelation, and heteroskedasticity. The template matrix X must be in maximum rank for regular lesser square estimation methods. Multi-linearity is caused by a correlation between two or more predictor variables.

- With multiple independent variables, a list of the most relevant independent variables uses specific approaches, like forwarding selection, backward exclusion, and incremental approach.

Logistic Regression

Logistic regression (LoR) is applied to find the probability of a reliant variable with two possible values, such as win/lose, pass/fail, alive/dead, or healthy/sick. In other words, it is used when the dependent variable is naturally a binary (0/1, True/False, Yes/No) variable. There is, the value of Y is 0 or 1, and it represents a binomial distribution. Logistic regression is a predictive analysis and is widely applied to explain data and clarify the connection between one dependent binary parameter and one or more forecasters.

The following equation can be used to estimate the probability of remaining cases based on independent parameter values (forecasters):

$$\text{odds} = \frac{p}{1-p} \tag{5.16}$$

where p is the probability of event occurrence, $1-p$ is the probability of not event occurrence. Then, applying the logarithmic function, we get

$$l = \ln(\text{odds}) = \text{logit}(p) \tag{5.17}$$

which is the same as applying a logistic function top? The logit function plays the link function role. Considering independent (X) and dependent (Y) variables,

$$\text{logit } E(Y) = Xa \tag{5.18}$$

where Y is a vector of dependent variables, y_i ($i = 1..., n$), X is the matrix of independent variables, a matrix of row-vectors x_i or n-dimensional column-vectors X_j, and a is the $(n+1)$-dimensional vector.

Because the odds are not a linear combination of the predictors, logistic regression is a nonlinear model. There is no precondition for the linear relation between the dependent and the separate variables.

There are three different Logistic regression models.

- **Binomial** – the outcome of the dependent variable only has two possible types (0 or 1);
- **Multinomial** – the outcome of the predictor(s) has three or more possible types; and
- **Ordinal** – the predictors are ordered.

Because maximum probability estimates at low sample sizes are less efficient than ordinary sizes, the logistic regression must be considered before use.

Generalized Linear Model

A generalized linear model is a common linear method of regression in which the error distribution of dependent variables is not normal. The generalized linear model (GLM) uses a connection function to generalize linear regression. It also allows reliance on their expected value on the degree of variance of each calculation.

The linear model of regression provides a linear combination of the predictors with the expected value of a given instance. This shows that a continuous change in a predictor induces a continuous change in the output, and the model acts like a variable of a linear response. However, if the dependent variable has a distribution other than a normal distribution, this is not sufficient. A dependent variable's distribution function is seldom a normal distribution.

Utilizing a particular distribution in the exponential family, including regular, exponential, gamma, Poisson, Bernoulli, … the mean of the dependent parameter (X) is in connection with the independent variable (Y) is

$$E(Y) = f^{-1}(Xa) \tag{5.19}$$

where $E(Y)$ is the expected value of Y, a is the vector of model parameters, and f is the link function (from the exponential family).

The applying of the maximum likelihood approach, iteratively, estimates the model parameter, a.

Polynomial Regression

The regression equation is a polynomial regression (PR) equation if the independent variable's power is greater than one. Equation (5.20) represents a polynomial equation.

$$y = a + bX^2 \qquad (5.20)$$

The suitable line is not straight in the PR system. It is normally a curve in the data points (Figure 5.4).

Stepwise Regression

This kind of regression is used when dealing with several different variables. In this approach, a programmed procedure, which requires no human interference, is used to select independent variables.

This achievement is accomplished by using statistical values such as R-square, t-states, and AIC metrics to define critical parameters. Stepwise regression (SR) fits the regression model by adding/dropping covariates one at a time based on a specific standard. The commonly utilized SR techniques are as follows:

- Standard SR does two things. It adds and deletes predictors for each move.
- The selection of forwarding begins with the most relevant model predictor and then the addition of parameters for each move; and
- Backward removal commences with all model forecasters and removes for each step the least significant variable.

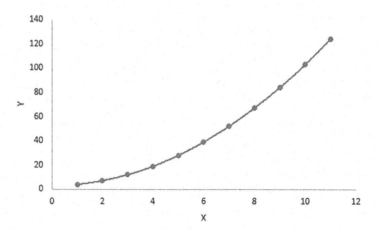

FIGURE 5.4
Polynomial regression.

The goal is to optimize prediction power with the minimum number of forecasting parameters.

Ridge Regression

Ridge regression (RR) is a methodology used when data are highly correlated with multicollinearities. Although the lowest square estimate is balanced in multi-collinearity, their variances are significant, which make the data very different from the true one. When adding some partiality to the regression results, the RR eliminates defections.

The RR resolves the multicollinearity problem through shrinkage parameter λ (Equation 5.21).

$$= \text{argmin} \ \|y - X\beta\|_2^2 + \lambda \|\beta\|_2^2 \tag{5.21}$$

In this equation, there are two components. The first one is the least-square term, and the second one is λ of the summation of β^2, where β is the coefficient. This is added to the least square term to shrink the parameter to have a very low variance.

Lasso Regression

Like the Ridge regression, it penalizes the absolute scale of the regression coefficients in Lasso (Last Absolute Shrinkage and Selection Operator). Besides, the volatility and exactness of the linear regression models can be minimized (Equation 5.22).

$$= \text{argmin} \ \|y - X\beta\|_2^2 + \lambda \|\beta\|_1 \tag{5.22}$$

Lasso regression (LR) differs from RR in the way that it applies Absolute Values (Avs) in the Penalty Function (PF), as an alternative of squares. This technique leads to penalizing values that cause several of the variable assessments to turn out to be precisely zero. Higher the penalty used, the further the estimates get shrunk towards absolute zero. This results in parameter selection out of given n variables.

Elastic Net Regression

Elastic Net is a hybrid of Lasso and Ridge Regression techniques. It is trained with L_1 and L_2 before regularization. Elastic-net is useful when multiple features are connected. Lasso is to be expected to pick one of these at random, while elastic-net is likely to pick both.

$$\hat{\beta} = \text{argmin} \left(\|y - X\beta\|^2 + \lambda_2 \|\beta\|^2 + \lambda_1 \|\beta\|_1 \right) \tag{5.23}$$

One functional advantage of Lasso-Ridge exchange is that Elastic-Net is allowed to inherit some of the stability of Ridge when it is rotated.

Many types of regression models are available, and it is crucial to choose the most suitable technique based on independent and dependent variables, data dimensions, and other necessary data features. The following are the main factors to consider when choosing the right model of regression:

- Data exploration is an essential component of a predictive model. Before choosing the right strategy, the first step should be to understand the relation and impact of parameters.

- To evaluate the benefit of fit for different models, we can analyze different metrics like the statistical importance of variables, R-square, Adjusted r-square, AIC, BIC, and the error term. Another one is Mallow's Cp criterion. This essentially checks for possible bias in the model by evaluating the technique with all potential submethods.

- Cross-validation is the best method to estimate techniques applied for prediction. There are datasets that are divided into two groups: data for training and data for validation. In this approach, a simple mean squared difference between the actual and predicted values gives an evaluation for the prediction accuracy.

- If the data set has multiple confounding variables, it should not be chosen as the automatic model selection method.

- It will also depend on the objective. A less powerful model may be easy to implement as compared to a highly statistically significant model.

- Regression regularization methods work well when the variables in the data set have high dimensionality and multilinearity.

Time Series Forecasting

A time series (TS) is a string of data points, temporally indexed for the time of observation, and has a first time to order (Daily, Monthly, Quarterly, and or Yearly) [32]. In other words, a TS is a sequence of measurements over time, usually collected at equally spaced intervals. Hence, it can be considered a sequence of discrete-time data [33].

Time series are widely used in different domains such as statistics, signal processing, pattern recognition, econometrics, finance, supply chain management, production planning, weather forecasting, and in any subject of applied science and engineering that includes discrete-time measurements. It has a well-established theoretical basis in statistics and dynamic systems theory.

Residual Pattern

TS analysis is an analysis domain that comprises methods and algorithms to study and analyze time-series data [34]. The main aim of TS analysis is to extract descriptive statistics and insights from data. In a TS, findings closely linked in time will be more closely linked than statements that occur a bit earlier or later. The amount of the relationship can be measured utilizing a stochastic model for a TS. On the other hand, TS forecasting is the use of a model to forecast future values based on observed values.

There is an original tread in TS analysis, which is as follows:

- Plotting the TS on a chart to identify its features, involving the presence of trend, seasonal and cyclical variation, as well as a residual pattern. This is then followed by determining outliers or anything like changes concerning components of TS.

- Next comes eliminating the trend or seasonal features of the signal. This would be performed using either differencing or using any transform function, such as exponential. The transformed signal must be stationary. If a TS is stationary, its statistical properties such as average and variance do not change over time. To build a predictive model for TS, this is a necessary condition. Also, the mathematical complexity of the predictive model decreases with the stationarity assumption.

- Then is the development of a predicting model for the stationary transformed signal. Please note that there are several algorithms to utilize for the forecasting model. Also, further analysis is required to determine the best approach to deploy, according to the data characteristics. The principle of parsimony should be considered while developing a TS forecasting model. Considering the principle of parsimony, the model with the smallest possible number of parameters which provides an adequate representation of the underlying data is always the best model. The principle of parsimony must be considered in all modeling problems.

- Validation of the selected model (or models) to find out the performance of the model and forecasting accuracy. The validation process may consist of some split-sample or cross-validation processes. The aim is to choose the best technique for forecasting, considering the principle of parsimony and performance of the model.

- After the development of the forecasting model and generation of the forecasted values, it is essential to use the inverse transformation function (the transformation function that is used to make the TS signal stationary). So, the forecasted values will be on the scale of the original data.

- Utilize and implement a process to monitor the performance of the forecasting model over time. This process is usually performed by evaluating the stream of forecast errors that have occurred.

Several different models can be utilized to develop a forecasting model of a TS. For example, there are many approaches to modeling and forecasting trends. Therefore, selecting an appropriate predicting model is of heuristic significance.

Overfitting should be considered during model development. In this case, the accuracy of the forecasting model is very high for historical data. However, its performance is low when used for future TS data.

There are several methods for predictive modeling of TS. In this section, some of the most popular methods of forecasting deployed for TS forecasting are described, including exponential smoothing, normal self-representation moving (autoregressive moving average, ARMA), and integrated auto-repressive moving average (autoregressive integrated moving average, ARIMA).

Exponential Smoothing Models

In general, component data must be set to signal and noise. The signal is the pattern from the intrinsic dynamics of the data collection system. Smoothing can be seen as a strategy to differentiate between signal and noise, as much as possible. The function of a filter in estimating the signal from the data is thus smoother [35].

The simplest method of smoothing a TS is to use the averaging process on the current and the previous observations. In other words, the current observation is replaced with the average of observations (current and previous).

In most cases, the recent values of the observation affect the benefits of future observations much more in contrast to the older ones. Exponentially weighted smoother reacts faster to TS oscillations by multiplying geometrically decreasing weights to the previous observations. Therefore, the addition of a parameter is an exponentially weighted smoother λ as

$$\hat{y}_t = \lambda \sum_{t=0}^{T-1} (1-\lambda)^t y_{T-t} \qquad (5.24)$$

where $\hat{y}_t$ is the exponentially weighted average of the current and $T-1$ previous observations, and λ is the discount factor and $|\lambda| < 1$.

Exponential smoothers are utilized in estimating the constant and linear trend of a TS. For the constant trend, the first-order or straightforward exponential smoother is deployed, and for estimation of the linear trend, the second-order exponential smoother is utilized.

Using the exponential smoothing model, the prediction of the τ-step ahead observation is equal to the current value of the exponential smoother for a TS with a constant trend

$$\hat{y}_{T+\tau} = \hat{y}_T \qquad (5.25)$$

The prediction error is a function of λ, while larger values of λ result in larger values of error. A large λ may be due to a fast reaction to the forecast error, but it may also make the forecast model react faster to random fluctuations. The choice of the λ parameter is very crucial and can be estimated using some of the squared forecast errors. The λ that produces the smallest sum of the squared forecast errors is the best choice.

ARMA models

Deployed models for TS have many variations with regard to different stochastic processes. There are two primary and broadly used linear TS models: Autoregressive (AR) and Moving Average (MA) models. Combining these two gives the Autoregressive Moving Average (ARMA) and Autoregressive Integrated Moving Average (ARIMA) models [36].

An ARMA (p, q) model consists of two models: the AR(p) and MA(q) models. Utilizing ARMA models for univariate TS modeling leads to a usually satisfactory result. In an AR(p) model, the potential amount of the observation is supposed to be a linear arrangement of p past observations and a random error in addition to a constant term [37]. Mathematically, the AR(p) model is represented by:

$$y_t = c + \sum_{i=1}^{p} \varphi_i y_{t-i} + \varepsilon_t \tag{5.26}$$

where c is a constant term, y_{t-i} is the observed value at time $t - i$, ε_t is the random error at time t, φ_i is the model parameter, and i is the order of the model.

The AR(p) model regresses against past values of the TS. Alternatively, an MA(q) model uses past errors as the explanatory variables. The MA(q) model is expressed by

$$y_t = \mu + \sum_{j=1}^{q} \theta_j \varepsilon_{t-j} + \varepsilon_t \tag{5.27}$$

where μ is the average value of the TS, θ_j is the model parameter, and q is the order of the model.

The random error is considered as a Gaussian distribution. So, an MA model is a linear regression of the current observation of the TS against the random error of one or more prior observations. Since the random errors in the MA model cannot be neglected, fitting an MA model to a TS is more complicated.

A combination of AR and MA models leads to a general and useful type of TS model, known as the ARMA model. ARMA (p, q) model can be stated as

$$y_t = c + \varepsilon_t + \sum_{i=1}^{p} \varphi_i y_{t-i} + \sum_{j=1}^{q} \theta_j \varepsilon_{t-j} \tag{5.28}$$

where p corresponds to the order and q is the direction for the average model to pass.

It is essential that to use ARMA models, TS should be stationary or should be transformed into a stationary signal using a transformation function.

To determine the orders of the ARMA model for TS data, it is necessary to perform the autocorrelation function (ACF) and partial autocorrelation function (PACF). These statistical measures express the relationship between the various observations in TS. For modeling and forecasting purposes, it is often useful to plot the ACF and PACF against a sequence of time lags. These plots help in estimating the order of AR and MA models.

Note that the ACF and PACF of an ARMA (p, q) both exhibit exponential decay as well as damped sinusoid patterns, which makes the identification of the order of the ARMA (p, q) model relatively more complicated. For that, additional sample functions such as the Extended Sample ACF, the Generalized Sample PACF, the Inverse ACF, and canonical correlations can be utilized.

The ARMA models can only be deployed for stationary TS data. For nonstationary TS, ARIMA models can be used. The nature of many TS data, such as trading or weather signals, is nonstationary. This involves the TS, which contains trends and seasonal patterns. So, ARMA models are practically inadequate to adequately describe nonstationary TS data, which are frequently encountered in practice. For this reason, the ARIMA model is proposed, which is a generalization of an ARMA model to involve the nonstationarity case as well.

ARIMA Models

ARIMA model and its different forms are based on the well-known Box–Jenkins principle, and so these are also widely known as the Box–Jenkins models [38].

A nonstationary TS is transformed to stationary by applying finite differencing of the data points in ARIMA models. Mathematically, the ARIMA (p, d, q) model using lag polynomials is written as

$$\left(1 - \sum_{i=1}^{p} \varphi_i B^i\right)(1 - B)^d y_t = \delta + \sum_{j=1}^{q} \theta_j B^j \varepsilon_t \tag{5.29}$$

where p, d, and q are ARIMA model orders for autoregressive, integrated, and moving average components, respectively, and are integer values ≥ 0. Parameter d determines the order of difference. Often, $d = 1$ is enough to make the TS stationary. When $d = 0$, then the ARIMA model reduces to an ARMA (p, q) model.

An ARIMA$(p, 0, 0)$ is the AR(p) model and ARIMA$(0,0,q)$ is the MA(q) model. ARIMA $(0,1,0)$ is a special one and known as the Random Walk model. In trading and price series, the Random Walk model is a well-known model for forecasting.

Machine Learning Predictive Models

Machine Learning (ML) approaches can be utilized as a regressor for prediction modeling. This section introduces the support vector machine and artificial neural networks and explains their prediction ability.

Support Vector Machine and AVM for Support Vector Regression (SVR)

In 1995, support vector machine (SVM) was developed to address problems of pattern recognition and classification, including facial identification and text grading [39]. Nevertheless, wide applications in other fields, such as feature rough-up, regression estimates, and TS predictions, were soon found.

By choosing any specific part of training data known as support vectors, SVM seeks to find a decision rule with little generalization. In this approach, after nonlinear mapping input space into a higher-dimensional space, an optimal separating hyper-plane is constructed. Hence, the input space is not the most effective parameter on the quality and complexity of the SVM result.

SVM's teaching is equivalent to the solution of a linear quadratic programming problem that is a major aspect of SVM. Therefore, the SVM solution is always special and globally optimal in contrast to other methods of preparation. One major demerit of SVM, however, is that the scale of the instruction, which raises the computational cost, is enormous.

Vapnik has derived a general algorithm to use SVM for regression. The optimal decision hyperplane of SVR can be expressed by

$$y(x) = \sum_{i=1}^{N_s} \left(\alpha_i - \alpha_i^* \right) k(\mathbf{x}, \mathbf{x}_i) + b_{\text{opt}} \tag{5.30}$$

where N_s is the total number of support vectors, k is a kernel function, b_{opt} is the optimal bias of the hyperplane, $\mathbf{x}$ is the observed dataset, and $\mathbf{x}_i$ is the ith observed data and optimized Lagrange multipliers $\alpha = (\alpha_1, \alpha_2, ..., \alpha_N)^T$ and $\alpha_i \geq 0$.

The Kernel is a nonlinear mapping for mapping the input space dots into a sizeable dimensional space. In this high-dimensional space, the optimal separating hyper-plane is built. This method also solves the problem when a linear decision is not possible to separate training points.

Some of the accessible kernel functions are

- Linear kernel $k(x, y) = x^T y$
- Polynomial kernel $k(x, y) = \left(a + x^T y \right)^d$
- Radial basis function (RBF) kernel $k(x, y) = e^{\frac{-\|x-y\|^2}{2\sigma^2}}$
- Neural network kernel $k(x, y) = \tanh\left(ax^T y + b \right)$ where a and b are constants.

Artificial Neural Networks

The model of artificial neural networks (ANNs) was proposed as a first prediction and simulation technique and has been used frequently as a modeling and prediction tool in recent years [40]. ANNs' primary aim was to create a model for the simulation of human brain behavior. The human intellect handles data in different ways than traditional digital computers [41].

The neuron is the brain's central structural portion and information processing module. The human brain is a very efficient structure for processing, learning, and reasoning information, as millions of neurons are present [42].

While ANNs were mainly bio-motivated, they were then applied in several different fields, including prediction and classification problems. The overall structure is a black-box model type often used for modeling a nonlinear, high-dimensional dataset. Hence, it does not require the development of analytical forecasting procedures or underlying knowledge of the system.

The main features of an artificial neural network that make it ideal for forecasting purpose are:

- They are inherently data-driven and self-adaptive. So, it does not need to determine a particular model form or to provide any a priori assumption about the data.

- ANNs are naturally nonlinear. Therefore, they are more practical and accurate in modeling and forecasting linear and nonlinear data.

- ANNs are universal functional approximators. ANNs can estimate any uninterrupted function to any desired performance and fit many historical observations accurately.

- ANNs can deal with different data problems, such as erroneous, defective, or fuzzy data.

Multilayer perceptrons are the most broadly used ANNs for forecasting purposes, which may have one or several hidden layers. Multilayer feedforward ANNs are multivariate statistical models deployed to relate n independent (predictor) variables to one or more dependent variables. ANNs have three distinct layers: the input layer, which is composed of the original predictors, the hidden layer that comprised of a set of interface variables, and an output layer that produces the responses. Weighted links connect these layers. Each variable in a layer is known as a node. Figure 5.5 illustrates a typical three-layer ANN structure.

A linear combination of the inputs to a node in any layer is mapped by an activation function to the output of the node. This output is one of the inputs for one- or more-layer nodes. The activation functions are usually nonlinear and are selected from a broad range of functions, including sigmoidal, linear, hyperbolic tangent, and Gaussian.

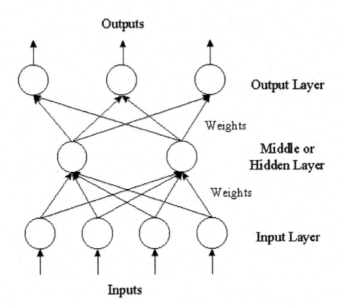

FIGURE 5.5
Structure of a typical ANN [40].

The linear combination of inputs to a node in the hidden layer can be written as

$$z_u = \sum_{j=1}^{n} w_{1ju} x_j + b_u \tag{5.31}$$

where z_u is the output of node u at first hidden layer, w_{1ju} are the weights of links and act as unknown parameters that should be estimated, x_j are the input parameter, and b_u is the bias parameter of the node. The activation function is applied to generate the output of each node in all layers.

The parameters in the underlying model, including weights and biases, must be estimated by the training procedure. Several methods of training were used for parameter estimation by reducing the total number of quadrupled errors over all the ANN outputs. Learning rules are known as optimization techniques used to mitigate error. Backpropagation and profound learning are the most common rules.

Overfitting is a common issue with ANNS. To tackle this problem, there are some ways, such as the following:

- Lower the number of parameters, including the number of hidden layers in every layer,
- Stopping the parameter estimation process after a limited number of iterations,

- Deploying cross-validation to determine the number of iterations, and
- Considering a penalty function to the error function.

When using ANN for prediction problems, the selection of a suitable network structure is important. In order to determine the number of hidden layers, multiple nodes in each layer, and the choice of the activation function, a parsimony principle should be considered.

In addition, it is often necessary to properly transform or rescale training data to achieve the best possible results. All input variables are commonly standardized or resized, so small random values are produced for the weight and bias starting values.

Summary

This chapter explained some popular toolsets that are used to complete the data analytics in industries, especially in mining. All toolsets were divided into three main groups, including the statistical approaches, state-space approaches, and predictive models. The selection of statistical tools and the analysis of variance were explained when there is this opportunity to use the correlation analysis. The state-space modeling and forecasting were individually investigated, and some details about the predictive techniques were explained. The investigated predictive models in this chapter included the regression, time series, and machine learning methods. This chapter tried to make a clear vision for the researchers to have better ideas in order to help them decide to select the best analytical toolset.

References

1. Gu, C., *Smoothing Spline ANOVA Models*. Vol. 297. 2013: Springer Science & Business Media, Berlin.
2. Hoaglin, D.C. and R.E. Welsch, The hat matrix in regression and ANOVA. *The American Statistician*, 1978. 32(1): pp. 17–22.
3. Rutherford, A., *Introducing ANOVA and ANCOVA: A GLM Approach*. 2001: Sage, Thousand Oaks, CA.
4. Goldberg, D.E. and S.M. Scheiner, ANOVA and ANCOVA: Field competition experiments. *Design and Analysis of Ecological Experiments*, 2001. 2: pp. 69–93.
5. Huberty, C.J. and S. Olejnik, *Applied MANOVA and Discriminant Analysis*. Vol. 498. 2006: John Wiley & Sons, Hoboken, NJ.

6. Young, M., G. Walton, and E. Holley, Investigation of factors influencing roof stability at a Western US longwall coal mine. *International Journal of Mining Science and Technology*, 2019. 29(1): pp. 139–143.

7. Onder, M. and E. Yigit, Assessment of respirable dust exposures in an opencast coal mine. *Environmental Monitoring and Assessment*, 2009. 152(1–4): p. 393.

8. Bouzalakos, S., A. Dudeney, and B. Chan, Formulating and optimising the compressive strength of controlled low-strength materials containing mine tailings by mixture design and response surface methods. *Minerals Engineering*, 2013. 53: pp. 48–56.

9. Thompson, B., Canonical correlation analysis. In Everitt, B., Howell, D.C. (eds.) *Encyclopedia of Statistics in Behavioral Science*, 2005: Wiley, West Sussex.

10. Cohen, P., S.G. West, and L.S. Aiken, *Applied Multiple Regression/Correlation Analysis for the Behavioral Sciences*. 2014: Psychology Press, Hove.

11. Benesty, J., et al., Pearson correlation coefficient. In *Noise Reduction in Speech Processing*. 2009: Springer, Berlin. pp. 1–4.

12. Abdi, H., The Kendall rank correlation coefficient. In Salkind, N.J. (ed.) *Encyclopedia of Measurement and Statistics*. 2007: Sage, Thousand Oaks, CA. pp. 508–510.

13. McLeod, A.I., Kendall rank correlation and Mann-Kendall trend test. R Package Kendall, 2005.

14. Zar, J.H., Spearman rank correlation. *Encyclopedia of Biostatistics*, 2005. 7: pp. 67–72.

15. Colwell, D. and J. Gillett, Spearman versus kendall. *The Mathematical Gazette*, 1982. 66(438): pp. 307–309.

16. Szmidt, E. and J. Kacprzyk, The Spearman and Kendall rank correlation coefficients between intuitionistic fuzzy sets. *In Proceedings of the 7th Conference of the European Society for Fuzzy Logic and Technology*, 18–22 July, Aix-les-Bains, France. 2011: Atlantis Press.

17. Tate, R.F., Correlation between a discrete and a continuous variable. Point-biserial correlation. *The Annals of Mathematical Statistics*, 1954. 25(3): pp. 603–607.

18. Varma, S., *Preliminary Item Statistics Using Point-Biserial Correlation and p-Values*. 2006: Educational Data Systems Inc., Morgan Hill, CA. Retrieved 16(07).

19. Häne, B.G., K. Jäger, and H.G. Drexler, The Pearson product-moment correlation coefficient is better suited for identification of DNA fingerprint profiles than band matching algorithms. *Electrophoresis*, 1993. 14(1): pp. 967–972.

20. Derrick, T.R., B.T. Bates, and J.S. Dufek, Evaluation of time-series data sets using the Pearson product-moment correlation coefficient. *Medicine and Science in Sports and Exercise*, 1994. 26(7): pp. 919–928.

21. Dziuban, C.D. and E.C. Shirkey, When is a correlation matrix appropriate for factor analysis? Some decision rules. *Psychological Bulletin*, 1974. 81(6): p. 358.

22. Croux, C. and G. Haesbroeck, Principal component analysis based on robust estimators of the covariance or correlation matrix: Influence functions and efficiencies. *Biometrika*, 2000. 87(3): pp. 603–618.

23. Rinne, H., *The Weibull Distribution: A Handbook*. 2009: CRC Press, Boca Raton, FL.

24. Liu, C., R.W. White, and S. Dumais. Understanding web browsing behaviors through Weibull analysis of dwell time. In *Proceedings of the 33rd International ACM SIGIR Conference on Research and Development in Information Retrieval*. 2010. ACM.

25. Tang, C., et al., Coupled analysis of flow, stress and damage (FSD) in rock failure. *International Journal of Rock Mechanics and Mining Sciences*, 2002. 39(4): pp. 477–489.
26. Klett, J., *Applied Multivariate Analysis*. 1972: McGraw-Hill, New York.
27. Martin, N. and H. Maes, *Multivariate Analysis*. 1979: Academic Press, London.
28. Hyndman, R., et al., *Forecasting with Exponential Smoothing: The State Space Approach*. 2008: Springer Science & Business Media, Berlin.
29. Hyndman, R.J., et al., A state space framework for automatic forecasting using exponential smoothing methods. *International Journal of Forecasting*, 2002. 18(3): pp. 439–454.
30. Bates, D.M. and D.G. Watts, *Nonlinear Regression Analysis and Its Applications*. Vol. 2. 1988: Wiley, New York.
31. Kleinbaum, D.G., et al., *Applied Regression Analysis and Other Multivariable Methods*. Vol. 601. 1988: Duxbury Press, Belmont, CA.
32. Chatfield, C., *Time-Series Forecasting*. 2000: Chapman and Hall/CRC, Boca Raton, FL.
33. Bowerman, B.L. and R.T. O'Connell, *Time Series and Forecasting*. 1979: Duxbury Press, North Scituate, MA.
34. Gasser, T., L. Sroka, and C. Jennen-Steinmetz, Residual variance and residual pattern in nonlinear regression. *Biometrika*, 1986. 73(3): pp. 625–633.
35. Gardner Jr, E.S., Exponential smoothing: The state of the art. *Journal of Forecasting*, 1985. 4(1): pp. 1–28.
36. Said, S.E. and D.A. Dickey, Testing for unit roots in autoregressive-moving average models of unknown order. *Biometrika*, 1984. 71(3): pp. 599–607.
37. Hannan, E.J. and J. Rissanen, Recursive estimation of mixed autoregressive-moving average order. *Biometrika*, 1982. 69(1): pp. 81–94.
38. Jenkins, G., *Autoregressive–Integrated Moving Average (ARIMA) Models*. Wiley StatsRef: Statistics Reference Online, 2014.
39. Balabin, R.M. and E.I. Lomakina, Support vector machine regression (SVR/LS-SVM)—an alternative to neural networks (ANN) for analytical chemistry? Comparison of nonlinear methods on near infrared (NIR) spectroscopy data. *Analyst*, 2011. 138: pp. 1703–1712.
40. Soofastaei, A., et al., Development of a multi-layer perceptron artificial neural network model to determine haul trucks energy consumption. *International Journal of Mining Science and Technology*, 2016. 26(2): pp. 285–293.
41. LeCun, Y., et al., Neural networks-tricks of the trade. *Springer Lecture Notes in Computer Sciences*, 1998. 2: pp. 112–121.
42. Rodriguez, J., et al., The use of artificial neural network (ANN) for modeling the useful life of the failure assessment in blades of steam turbines. *Engineering Failure Analysis*, 2013. 35: pp. 562–575.

6

Process Analytics

Paulo Martins and Ali Soofastaei

Process Analytics

Every organization has to implement and manage several processes in order to deliver a service or a product to customers. Process analytics comprises a collection of methods and tools used to analyze all relevant data produced during the process and operational execution. The objective behind process analytics is to support smart decision-making and process optimization and provide feedback on the efficacy and performance of the company's processes. When decision-makers receive performance analytics insights, they can quickly assess the impact of process management decisions on process targets and quickly respond to them in the form of control and improvement (or re-engineering) of operations [1–3].

Depending on the organization's IT infrastructure, the analytics of process activities can be performed in different systems such as a Business Performance Management (BPM) system, an Electronic Content Management system, an Enterprise Resource Planning platform, a Consumer Relationship Management system, a Manufacturing Execution System, and a Supervisory Control systems, among others. All these systems may be fed by multiple processes with massive data. Nevertheless, companies will continue to keep such information stored for use regularly during the decision-making process without appropriate analytical tools and efficient computing capable of extracting insights from the data.

The astonishing growth of Big Data capabilities has motivated companies to embed sophisticated analytics approaches to drive process management. Business Process Management has not kept up to date and often only relied on conventional modeling methods [4–10]. In this new scenario, the application of Advanced Analytics technologies and methods can significantly enhance business performance and efficiency, providing standard tools to support, diagnose, and optimize processes.

In order to improve the performance of the company, big data analysis was believed to be a primary feature. In order for BDA to be operational, the analysis would operate with the processes. Process analytics and BDA yield intelligent business process management. By deploying real-time intelligence and Advanced Analytics directly linked with the automated process, an organization can identify trends and gaps and monitor results. Figure 6.1 shows an overview of the main areas where processes analytics can have a significant impact.

Process Analytics Tools and Methods

The continuous search for process improvement and operational excellence raises the necessity of process analytics methods and tolls inside an organization. Analysis of business processes is a method of natural management to ensure organizational goals and continuity. In order to understand process output effectively and establish management strategies, data plays a central role. It is not only necessary to understand past events but also to analyze the current behavior and predict what will happen in the future. In this section, two crucial operational performance fields that have their foundations based on the process analytics principles will be discussed: The Lean Six Sigma (LSS) and Business Process Analytics (BPA). The main objective is to describe the influence of digital technologies and Big Data elements and present how Big data analytics (BDA) has been applied to leverage impact on business processes.

Lean Six Sigma

The LSS methodology is a robust continuous improvement approach widely used in business that combines the concepts of Lean Manufacturing and Six Sigma. Lean manufacturing is a management methodology that originated in Japan and was consolidated inside Toyota Motor company, and it aims at doing the right things at the right time and in the right amount, reducing waste, being flexible, and being open to change. Six Sigma is a data-driven business strategy built under the DMAIC (Define, Measure, Analyze, Improve, and Control) problem-solving methodology that aims to implement high standards of operational practices to produce quasi-perfects products and services and reduce the production process variation to a microscopic level. It is, by definition, extremely analytical and profoundly rooted in statistical analysis [11–20].

Characteristics of LSS Data and Methods

The analysis of process data is inherent in all five phases of the DMAIC approach. However, special emphasis is dedicated to the analytical methods and tools used during the Measure and Analyze steps. In the Measure step, the focus is on (1) selecting the relevant output characteristics, (2) assessing the performance specifications, and (3) understanding current process capabilities. Then, the Analyze step is applied to (1) analyze the current process performance, (2) monitor the potential Critical to Process metrics, and

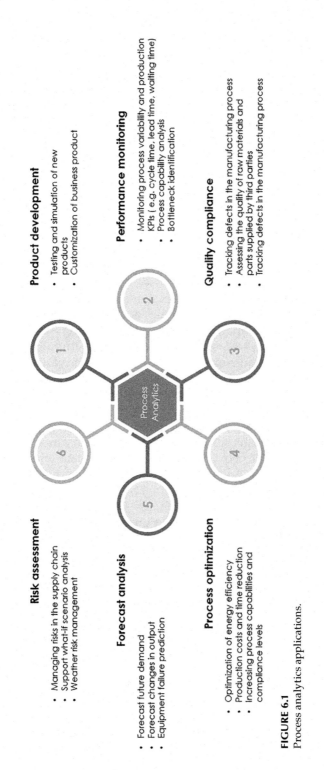

Product development

- Testing and simulation of new products
- Customization of business product

Performance monitoring

- Monitoring process variability and production KPIs (e.g. cycle time, lead time, waiting time)
- Process capability analysis
- Bottleneck identification

Quality compliance

- Tracking defects in the manufacturing process
- Assessing the quality of raw materials and parts supplied by third parties
- Tracking defects in the manufacturing process

Risk assessment

- Managing risks in the supply chain
- Support what-if scenario analysis
- Weather risk management

Forecast analysis

- Forecast future demand
- Forecast changes in output
- Equipment failure prediction

Process optimization

- Optimization of energy efficiency
- Production costs and time reduction
- Increasing process capabilities and compliance levels

FIGURE 6.1
Process analytics applications.

(3) identify what resources will be needed for improvement [16]. Although the analytical effort is more intense in these two phases, the entire DMAIC framework relies on data-driven techniques and methods.

Typically, the data collected in the LSS project consist of some process characteristics that should be improved (Y) and some influence factors (X). These data are used to compound plenty of analytical tools (maps, graphics, and statistics) such as Value Stream Map, time value map, histogram, box-plot Pareto charts, Analysis of Variance, process capability calculation, control charts, correlation matrix, simple linear regression, multiple regression, and Design of Experiments, among others.

Integration of Big Data and LSS

Although those traditional analytics tools and methods can be used to develop processes improvement initiatives, there are some limitations to be considered in the current data-intensive and high-technological scenario that shapes the industry 4.0. Some weaknesses are related to the excessive time required to manipulate and process massive and multivariate data and the incapacity of integrating multiple sources on a real-time basis.

A new generation of analytical tools is transforming the way companies uncover complex processes issues and inefficiencies through a combination of sophisticated techniques and methods to process the vast amount of available data. These techniques can be mixed with more traditional methods and in-depth comprehension of the physical context [8]. Since there is an explosion of available data, it is essential to consider the application of new mining techniques, such as BDA, data mining, and process mining. Big Data techniques (natural language processing, text mining, machine learning, deep learning, and artificial neural networks), primary data mining techniques (clustering, association, prediction, and classification), and process mining algorithms are revolutionizing the entire LSS cycle, becoming a crucial component to support smart business decisions. Indeed, data mining techniques and processing are a part of BDA [16]. Figure 6.2 shows some BDA techniques and technologies that can be embedded in the analytics cycle to ensure the continuous improvement of the flywheel.

In the presence of unstructured data such as text, video, and voice records, text mining, video mining, and natural processing language can support analysis to clarify the problem definition and help the scope be defined in the Define phase. During the Measure phase, the use of advanced statistical techniques is considered a more efficient approach to understanding a process behavior and identifying waste, unhidden bottlenecks, and operational rigidities. Furthermore, data mining techniques, artificial intelligence, and machine learning are commonly applied to prepare data and deal with data quality issues (missing, duplicated, and incorrect values) [23, 24].

In the Analyze phase, traditional LSS programs focus on simple exploratory data analytics techniques. Advanced methods such as association

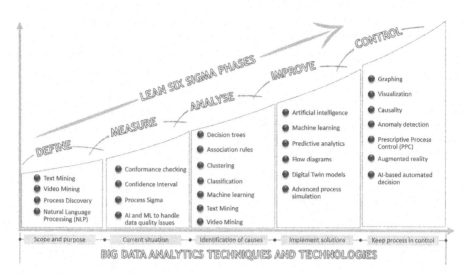

FIGURE 6.2
BDA techniques and technologies applied in the LSS cycle [16,17].

rules, clustering, and classification are commonly applied to find patterns in the process data. In some cases, machine learning algorithms are ideal for dealing with massive data without strict learning requirements [16]. By intelligence, they can quickly learn, evolve, and capture behaviors of the processes [4]. For example, a pharmaceutical company decided to discover the root causes of variability in a strategic production process applying neural networks (a machine learning technique) to model the relevant combinations and impact of the variables. After this first interaction, the five variables with the highest impact were determined, and the team then focused their optimization efforts to achieving a 30% improvement on yields [21]. Moreover, process discovery may be efficient with process-based data that contains a timestamp label.

The IMPROVE phase is supported by several BDA alternatives. Artificial intelligence algorithms have proved to be a powerful approach to investigate and define optimal process parameters. For example, genetic algorithms, a type of AI algorithm, has been widely applied for energy efficiency optimization.

In the last phase, that is Control, alarms can be created using unsupervised learning algorithms while conformance checking is applied to check the established model from event logs and physical processes [16]. One of the most applied methods to control process variability is the Statistical Process Control (SPC) represented by standard quality control charting techniques (e.g., Shewhart charts, X-bar and R charts, etc.). More than 50,000 SPC charts are popular for modern fabs, each of which tracks a specific quality variable [9]. However, control charts are limited in their explanatory power. In most of the cases, a univariate approach is used assuming a normal distribution (which may not correctly represent the process behavior). On the flip side, machine learning algorithms can overcome SPC performance by working with high-dimensional data and incorporating the Big data available in the processes operation to enhance the model's performance. Going one step further, sophisticated modeling techniques can be developed to recommend the best action, supporting frontline operators to keep the parameters under control. This new operational approach for process monitoring and improvement is named Prescriptive Process Control (PPC) [23].

The same positive effect can be observed when LSS is incorporated into Advanced Analytics projects. For example, a significant global financial firm used LSS methods to help its analytics projects by having a more structured and measured approach to perform analytics activities. Additionally, they discovered that analytical methods enabled improvement of other projects not directly linked to business analytics across the company [17]. In [18], the LSS methods were applied to improve in the Digital Curation lifecycle – selection, preservation, maintenance, collection and archiving of digital information – and they concluded that all activities may be executed and optimized by using the DMAIC methodology.

Big Data can solve the most complex causal inference problems, but it still faces the lack of a structured approach. LSS has a well-defined and robust structure but needs a refresh in order to reach a new level and meet modern processes requirements. These arguments make their combined use extremely promising. Companies with a stable LSS DNA, like GE, CISCO, and Intel, have successfully integrated BDA into their analytical initiatives [19,20].

According to [22], CRISP-DM is an industry-tested way to direct efforts in data mining and could be integrated into the DMAIC roadmap, providing an interactive structure between its phases. The combination of a problem-driven approach (DMAIC) and a data-driven approach (CRISP-DM) can help both fields be more valuable. The authors suggested the integration of data analytics experts (data scientists) into the organizational structure of LSS teams, with the intent of performing complex analyses and translating data science conclusions to business conclusions.

Therefore, as the modern processes are becoming smarter by incorporating various technologies, the flood of data must be embedded in the LSS cycle. Since BDA aims to deliver more profound knowledge into processes data that previously were not accessible due to technology and methodology limitations, it can cohesively work together with LSS methods, passing insights and wisdom back and forth like old friends.

Business Process Analytics

Business Process (BP) can be defined to achieve a business objective or objectives through organized tasks or activities carried out manually or automatically [25]. BPM is an operations management field dedicated to overseeing the execution of BPs to ensure consistent results and identify and implement improvement opportunities [6]. Process Analytics is an essential component of the BPM discipline that incorporates various data manipulation activities – from data acquisition, discovering, modeling, and analysis to delivering – applied to process execution data in order to deliver insights and support smart data-driven decisions [12].

Over the last decades, many companies have embraced a process-oriented approach in order to improve process quality and efficiency. The different types of data (structured, semi-structured, and unstructured) explained in Chapter 3 are commonly found in contemporary organizations' processes. Data form an inseparable part of business processes and are derived from the execution of processes, documentation, and definition of processes, process models, process variants, business process objects, and data generated or exchanged during process operation [12]. Furthermore, it enables businesses to automatedly collect information to understand process performance, cost drivers, and risk triggers by replacing paper and manual processes with software. Real-time digital process performance reports and dashboards allow managers to address problems before they become critical.

Characteristics of Process Data and Analytics Structure

In [26], industrial processes were compared to a two-sided coin, where the first principles are the model and where uncertainty is on the other side. Uncertainty is a pervasive characteristic of process operations and control, presenting a time-varying and dynamic behavior. There are multiple process measurement technologies available to collect data from conventional sensors, images, and videos. The assessment of process performance is a multi-dimensional task because it cannot be calculated using only one indicator, such as productivity. It results from several distinct parameters, measures, and performance [26]. The following desirable features are required for multi-level sophisticated process data analytics:

1. Scalable and interpretable (up to thousands of variables)
2. Incorporate and use all types of data (e.g., processing data, spectra, vibration, and image data)
3. Compared to other approaches, it is relatively easy to apply to the real process
4. Real-time control and decision-making online technology
5. Offline issues resolution as a valuable tool for constant improvement [27]

In the last section, control charts were described as a process monitoring technique widely used for process stability control in the LSS context. Similarly, the BPM discipline applies Process Analytics methods for monitoring industrial process performance and execution of data analytics tasks. The following steps are considered:

1. Collection of data that represent the operational functions of the process, followed by an extensive data cleaning procedure.
2. Fault detection techniques, such as linear differential analysis, may be used in the presence of many defective data.
3. Design of standard data analysis models combined with defect detection indices and control limits.
4. Conduct fault analysis and problem solving.

Most of the current work on process analytics is dedicated to exploring new knowledge and analysis including (1) trend analysis, using methods to investigate data and monitor business opportunities, (2) what-if assessment, analyzing different scenarios to reach the best business decision, and (3) advanced analysis, applying sophisticated methods to identify patterns and relationships in strategic factors that affect the business. A typical BPA structure has its data sources integrated into data warehouse servers using Extract, Transform, and Load (ETL) or complex-event processing engines. The transformation of process data to process insight data is executed by multi-tier servers such as OLAP, enterprise search engines, data mining engines, text analytics engines, and reporting servers implemented at the top of the data warehouse. Finally, front-end applications and visual interfaces are deployed to facilitate the user's interaction and comprehension of analytics results [12].

Integration of BDA into Process Analytics Methods

The fourth industrial revolution is transforming the management of business practices, affecting the way business process is conceived, monitored, controlled, and improved. This new scenario creates a demand for complex, dynamic, and, often, knowledge-intensive solutions. Although significant improvements have been consolidated over the last years, BPM systems and BPA techniques present limited capacity to perform analysis in complex systems. There is an increasing necessity of integrating ad hoc and routine work, based on flexible workflows and task management [12].

Not suddenly, because of their exceptional ability to extract vast amounts of raw data and apply the best analytical techniques, BDA is considered a leading approach in the research process of big data. Automatic data analysis has become an important method used by many organizations, informing critical decisions previously based on decision-makers ' assumptions and expectations [11]. Unlike traditional analytics methods that focus on the cleanness of the data to avoid wrong decisions, BDA treats data errors and messiness as inherent to the process and applies robust models to extract features that overcome data quality issues [27].

As a critical factor in raising overall performance level and reduce performance variations, BDA can be integrated into the business process and shape a "new class of economic asset" and support high-performance companies to redesign their operations and outperform their competitors. Current software for BPA is idealized around two perspectives: performance and compliance [3,12]. The best-rated commercial solutions for process analytics such as IBM, Oracle, SAP, and ARIS combine process management/integration and business rule management software to assist business decisions [12].

However, traditional analytics tools and management practices present significant limitations to deal with a Big Data environment and the dynamic behavior of modern processes. Analytics applications depend on event data integration, but it is hard to obtain in a highly distributed environment. Moreover, the massive amount of data generated by distributed processes cannot be managed by traditional systems [1]. Current capabilities of OLAP technologies are not designed to extract patterns among process graph entities and perform analytics in multidimensional graph data, and considering different perspectives and granularities may be a challenging task [12]. Modern Big data technologies such as MapReduce and Hadoop, and high-level languages for data analysis, such as Pig, SCOPE, Sawzall, and Sphere, are available to deal with parallel processing issues.

In [10], the authors presented some Big Data systems-based solutions to support business process' data-intensive operations, showing how they can help data collection and provide the necessary infrastructure to analyze multiple data and generate insights more efficiently than current state-of-the-art process analytics techniques.

Nowadays, a broader spectrum of the application of BDA has been recognized by business leaders. These methods pervade different hierarchical levels in process industries, varying from passive applications (process monitoring and soft sensing) created to support frontline workers to oversee and manipulate process variations, to active applications (optimized control and complex decision-making) that directly influence process efficiency [28].

Machine learning supervised and unsupervised methods have been developed for plenty of process data applications designed for process monitoring, fault prediction, and diagnosis. Most recently, process monitoring methods have incorporated quality-relevant diagnosis based on supervised learning models for the diagnosis of undesirable conditions caused by unexpected process variations. This approach is referred to as Supervised Monitoring and Diagnosis, and the main objective is to monitor and diagnose quality problems before they occur [29].

The first application of deep learning techniques to soft sensing was reported in [28]. The study used a Deep Neural Network (DNN) to predict the cut-point temperature of heavy diesel in a crude distillation plant. The literature also contains relevant cases of application of DNNs for process monitoring and fault detection and diagnosing in big process data, with outstanding performance [30,31]. Predictive modeling using Support Vector Machine (SVM) and neural networks has been widely applied in optimization control projects to reduce the level of uncertainty that is intrinsic in the optimal control problem, resulting in improved control performance [28].

Reinforcement Learning, another machine learning technique, is used as a potential application for optimal process control when no model information is provided. The models can adapt to time-varying environments intrinsically and have a high potential for complex manufacturing processes [32,33].

Cases & Applications

Big Data Clustering for Process Control

Concerned with the rapid growth of data, the huge quantity of parameters to measured, and the dynamically changing environment in its factory, the Italian company Whirlpool used a Big Data clustering method to detect real-time variations in a manufacturing process of washing machines. The approach extended traditional clustering algorithms (like k-Means), enabling better comprehension of the nature of the process and more efficient Big Data processing. Furthermore, the developed model can perform root-causes analysis, providing insights regarding process wastes.

The primary goal was to define average values for three parameters (power, rotation speed, and total water inlet) and develop a solution able to identify anomalies in functional tests. The method used a combination of scalable K-means for initial interaction and K-medoids (another partitioning algorithm) enhanced by FAMES (FAst MEdoid Selection). The team has developed a solution for comparing the test series by their type and based on cluster tests. By performing a final comparison with standard samples, the model detects unusual data. The results confirmed the model's effectiveness to detect anomalies and identify specific problems based on the similarity with the cluster [9].

Cloud-Based Solution for Real-Time Process Analytics

Scenarios that involve high complexity and highly distributed process form a perfect condition for cloud computing applications. By having means for extracting and correlating process events, performance analytics can be provided in real time. In [34], cloud-based architecture is proposed to allow for continuous improvement of enterprise processes, measuring the performance of cross-functional activities at meager latency response rates. Using a set of Business Analytics Service Unit nodes and a Global Business Analytics Service component, local (inter-departmental) and global (cross-organizational) complex processes can be monitored and analyzed. With these features, the devised solution collects data generated from distributed heterogeneous systems, stores a massive amount of process data, and infers knowledge based on the acquired information.

The event repository uses a column-oriented NoSQL database (HBase) running on top of the Hadoop Distributed File System, presenting an outstanding performance for timely access to critical data due to its clustering capabilities and in-memory cache distribution. Furthermore, to ensure instantaneous identification of the sequence of events, an event-based model and an event correlation algorithm are implemented. Although the approach was not extended for mining processes and advanced analytics optimization techniques, the system presented an excellent performance for real-time activities monitoring and had the capacity of gathering distributed event logs regardless of operating system technology and location.

Advanced Analytics Approach for the Performance Gap

As a historical technology pioneer industry, Oil & Gas can obtain significant improvement from the application of Advanced Analytics. A McKinsey report estimates that AA and digital technology investments deliver returns as high as thirty to fifty times in a short-time period (few months) [5].

A typical control room crew of two or three operators on an offshore rig receives data from about 30.000 sensors and controls around 200 operating variables. Additionally, the team must consider the combination of these parameters and the effects of other factors that impact production, including weather conditions and wave heights. This high complex operational environment results in performance variations. Measurements of performance gap from an offshore field in the North Sea showed more than 5% of the difference between the highest- and lowest-performing control room teams. The company adopted AA methods to reduce production variations on a semi-submersible platform using a vast amount of structured and unstructured data extracted from 5,000 sensors and controls in three years. Applying machine learning, the analytics team could understand correlations and causalities and select the most relevant variables. Moreover, ML algorithms were also applied to identify and solve five process bottlenecks that were affecting production efficiency. Last but not least, failure prediction models were applied to reduce unplanned downtime caused by equipment failures, enhancing the reliability of critical processes, such as the gas compression system [5,8].

BDA and LSS for Environmental Performance

The literature contains plenty of studies providing clear evidence of how BDA increases resource efficiency in manufacturing plants [36]. Moreover, recent researches describe the potential benefits of BDA to evaluate and improve environmental performance (EP). At the same time, many other examples demonstrate the use of monitoring, visualization, and analysis methods and tools to minimize waste production, energy footprint, and enhance resource utilization [37–39].

Even though the integrated impact of BDA and LSS on environmental performance is not commonly explored in the literature, companies can improve their EP by adopting BDA to analyze massive environmental data created at multiple nodes in the organization and deliver insights to overcome the complexity surrounding environmental topics. LSS projects need BDA to manage the complexity of environmental metrics and accelerate LSS strategies to enhance EP. In [35], 201 manufacturing companies in North Africa familiar with BDA and LSS methods responded to a survey that reinforced the strong relationship between both disciplines in the development of EP initiatives. In [40], the authors explored the application of BDA, coining the term Sustainable Smart Manufacturing, and presented a set of BDA techniques and technologies applied for all phases in product lifecycle management.

Lead Time Prediction Using Machine Learning

Production planning and scheduling activities strictly rely on accurate prediction of process lead time (LT). However, most of the industrial practices estimate average LT based on historical data. This approach does not consider the influence of multiple factors in the lead time variability.

In [41], supervised Machine Learning (ML) was applied for LT prediction in a semiconductor manufacturing plant using historical data extracted from manufacturing execution systems. The analytics team selected forty-one features (thirty-five numerical and six categorical) considering their influence on the lead time and compared the accuracy results for eleven statistical learning models (three linear models and eight nonlinear models). The best result was achieved by the Random Forest method using the eight more relevant variables defined after a sensitivity analysis step.

The study reported in [42] investigated the prediction of order completion time by using real-time data collected from RFID job shop devices and employing a DNN approach. The DNN methods were outperformed by other advanced predictive models.

Applications in Mining

Tech businesses such as Google and Amazon are already gaining benefit from BDA innovations because they have less trouble handling Big Data in the process of innovation. Although the adoption of technology is relatively slow in the mining industry, operational challenges such as declining ore grades, deeper location of ore bodies, environmental and social constraints, and stakeholder pressures are demanding fast changes towards digital capabilities. The following topics explore the application of digital technologies, especially Advanced Analytics methods, in the mining process analytics, control, and improvement initiatives.

Mineral Process Analytics

One of the most promising fields for the application of advanced analytics in mining is mineral processing. Not surprisingly, the massive amount of real-time operational data is a rich source of value creation in the Big Data era.

The mining company Barrick Gold developed an integrated analytics platform using an operational process model applied in the management and optimization of daily operations and as a planning tool connected to the geology resource model. The company's main objective is to maximize the gold extraction based on an active mine to mill integration. The process model was built using a digital plant template configured with target data from plant schedule and process inputs. The digital template receives real-time data and transforms it into operational insights that are then used to develop predictive models. Operational data are classified in different operating models (e.g., running, down, idle, trouble) that provide the desired level of detail to explore improvement opportunities. Additionally, the team advocates that the higher quality of data obtained by the configuration of operating mode event frames allows the application of advanced analytics tools and methods. For example, to access a more reliable particle size value over time, a soft sensor was built using machine learning tools [49].

Another mining giant that demonstrates a digital footprint and has heavily invested in advanced analytics technology for mineral processing plants is Freeport-McMoRan. Its machine learning program is one of three major initiatives aimed to increase the company's copper production by 30%. Supported by the consulting company McKinsey, the team started to unlock insights and boost operational performance approaching data mining in an agile way. In the earlier stages, the team investigated data from the Bagdad plant mill, looking for patterns that revealed opportunities for improvements. Then, a machine learning model was employed to assess the actual mill performance against that expected by the local team. Before the project, the technical and operational team believed that there was only one type of ore feeding the plant, which resulted in a single procedure to adjust the mill's forty-two parameters. However, a mindset shift occurred when the mill's data were used to model the material feeding characteristics and appointed to seven different types of ore. Motivated by this initial finding, which suggested a 10% production improvement by adjusting mill's parameters based on the different ore types, an artificial intelligence model was developed to prescribe mill control settings according to the ore characteristics and plant equipment sensors, raising cooper production from that ore. In order to reach the next improvement level, the team introduced a new artificial intelligence algorithm to maximize cooper output [50].

Drill and Blast Analytics

Drilling and blasting are a crucial part of the mining value chain. The efficiency in these processes can improve loading and hauling operations and plant production and recovery performance, and reduce production and maintenance costs, and energy and water consumption. At Elkview, a Teck Resources operation, sensors installed on shovels measure the digging conditions, providing feedback for the technical team in order to optimize explosive distribution and quantity and reduce blasting costs and equipment wear. Consequently, shovels' productivity was maximized, and dust and vibration impacts were reduced. The team also gained further benefits through insights obtained from advanced visualization [43].

Another groundbreaking application of advanced analytics uses historical and current data from the entire mine value chain to create digital twin models. Machine algorithms are trained using geological block model data, various drilling and blasting parameters (drill patterns, explosive types, blast hole diameters, stemming height), and downstream process data (muck pile profile, digging rates, crusher, and mill throughput). In the software interface, engineers and geologists can simulate process performance by setting different design parameters [44].

Mine Fleet Analytics

Mining fleets are considered the most valuable source of data in a mining operation. Fleet Management Systems are globally employed to collect real-time data generated during the operational cycle, especially from the truck-shovel operation. The dynamic problem of allocation of trucks has continuously been researched over the last decades, and new advanced analytics approaches combined with modern positioning technology have recently emerged as an outstanding solution to deal with the high combinatorial and stochastic nature of dispatch decisions.

An important part of the management of truck-shovel performance, which is also an essential input for real-time allocation solutions, is the prediction of activity duration. The prediction, diagnosis, and optimization of mining charging and transportation using historical and reliable data were successfully applied with artificial intelligence and machine learning algorithms. [45–48].

In [45], the prediction of truck travel times (TTP) in open pit mine using machine learning was studied using data from Fushun West Mine in China. Common variables that affected TTP were classified into three groups: truck features, road features, and meteorological features (pressure, wind speed, temperature, relative humidity, and precipitation) and trained using three ML algorithms, that is, k-nearest neighbors (kNN), SVM, and random forest (RF). The prediction accuracy obtained from SVM and RF models outperformed the KNN and had 15.79% higher accuracy than that calculated by traditional methods.

The advanced analytics team of a big mining company in Brazil analyzed data from their truck-shovel system and identified that the surging of queues was the most counterproductive operational condition. Supported by the CRISP-DM methodology, the team explored possible causes. It implemented a data retrieval and preparation strategy to work with the relevant data, which were spread in four different data sources. The modeling phase consisted of the development of (1) an artificial neural network predictor, (2) a decision-maker module employed for generating alarms of the surge of the queue, and (3) an explainer interface designed to stratify the critical contributors for the triggered alarms and suggest proactive actions to prevent production losses. When implemented in a real production environment, the solution accounted for a 10% reduction of the queue time on the dumping areas (crushers) [46].

The abovementioned examples of prediction of mining activity duration present the potential of advanced analytics application for tremendously improving the performance of mining operations. Based on the relevant findings obtained by the recent use of evolutionary algorithms and machine learning methods, the critical challenge of real-time truck allocation will be solved soon.

Summary

Process analytics is an essential practice for companies in order to deliver high-standard services or products to their customers. Traditional analytics approaches that are fundamentally developed around process data, such as LSS and BPA, are facing several limitations when confronted with the challenges of the Big Data era, characterized by real-time, high-speed, dynamically changing, and multivariate requirements. Those methodologies can reach the next level by incorporating modern BDA techniques and technologies to boost their analytical power. Both the literature and industry are full of real case applications that support the introduction of BDA as game-changing technology in the process analytics and improvement cycle.

References

1. Vera-Baquero, A. and C. Ricardo, Business process analytics using a big data approach. *IT Professional*, 2013. 99(6): p. 1.
2. Wang, J., M. Yulin, Z. Laibin, G. Robert, and W. Dazhong, Deep learning for smart manufacturing: Methods and applications. *Journal of Manufacturing Systems*, 2018. 48: pp. 144–156.

3. Muehlen, M. and S. Robert, Business process analytics. In vom Brocke, J., Rosemann, M. (eds.) *Handbook on Business Process Management*. Vol. 2. 2015: Springer, Berlin.

4. Choi, T., W. Stein, and W. Yulan, Big Data analytics in operations management. 2018. doi: 10.1111/poms.12838.

5. Brun, A., T. Monica, and V. Thijs, Why Oil and Gas Companies Must Act on Analytics. 2017. [2020 accessed on 13/04/2020]; available from: https://www.mckinsey.com/industries/oil-and-gas/our-insights/why-oil-and-gas-companies-must-act-on-analytics.

6. Dumas, M., R. Marcello, M. Jan, and R. Hajo, *Fundamentals of Business Process Management*, Second Edition. 2018: Springer, Berlin.

7. Moef, A., P. Robert, L. Samir, T. Simon, and B. Rodolphe, The industrial management of SMEs in the era of industry 4.0. 2017.

8. Brun, A., Improving operations performance through artificial intelligence, digital, and advanced analytics applications. In *Society of Petroleum Engineers*, 23 – 25 October 2018 at the Royal International Convention Centre, Brisbane.

9. Stojanovic, N., D. Marki, and S. Ljiljana, Big data process analytics for continuous process improvement in manufacturing. In *IEEE International Conference on Big Data*, Oct 29–Nov 01, 2015, Santa Clara, CA.

10. Sakr, S., et al., Business process analytics and big data systems: A roadmap to bridge the gap. *IEEE Access*, 2018. 6: pp. 77308–77320.

11. Maroufkhani, P., et al., Big data analytics and firm performance: A systematic review. *Information*, 2019. 10(7): pp. 1–21.

12. Beheshti, S., et al., *Process Analytics: Concepts and Techniques for Querying and Analyzing Process Data*. 2016: Springer, Berlin.

13. Wamba, S. and M. Deepa, Big data integration with business processes: A literature review. *Business Process Management Journal*, 2017. 23(3): p. 16.

14. Chang, C., Y. Fan, H. Dexian, and L. Wenxiang, Data-driven soft sensor development based on deep learning technique. *Journal of Process Control*, 2014. 24: pp. 223–233.

15. Bass, I. and L. Barbara, *Lean Six Sigma using SigmaXL and Minitab*. 2009: McGraw-Hill, New York.

16. Dogan, O. and G. Omer, Data perspective of Lean Six Sigma in industry 4.0 era: A guide to improve quality. In *International Conference on Industrial Engineering and Operations Management*. March 6-8, 2018, Hilton Bandung Hotel, Bandung.

17. Fogarty, D., Lean six sigma and advanced analytics: Integrating complementary activities. *Global Journal of Advanced Research*, 2015. 2: pp. 472–480.

18. Arcidiacono, G., E. Luca, F. Fallucchi, and A. Pieroni, The use of Lean Six Sigma methodology in digital curation. 2017.

19. Forgaty, D., Lean six sigma and Big Data: continuing to innovate and optimize business processes. *Journal of Management and Innovation*, 2015. 1: pp. 2–20.

20. George, M., R. David, P. Mark, and M. John, *The Lean Six Sigma Pocket Toolbook*. 2005: McGraw-Hill Education, New York. (digital version not available-personal book)

21. Sony, M., A. Jiju, P. Sung, and M. Michael, Key criticisms of six sigma: A systematic literature review. 2018.

22. Zwetsloot, I., K. Alex, A. Thomas, and K. Henk de, Lean Six Sigma meets data science: Integrating two approaches based on three case-studies. 2018.

23. Schmitiz, M., Overcome SPC with Prescriptive Big Data Analytics. 2016 [cited 2020 11/04/2020]; available from: https://medium.com/@mSchmitz_/overcome-spc-with-prescriptive-big-data-analytics-d4f1d8cf2a78.

24. Gupta, S., M. Sachin, and G. Angappa, Big data in lean six sigma: A review and further research directions. *International Journal of Production Research*, 2019. doi: 10.1080/00207543.2019.1598599.

25. Motahari-Nezhad, H., S. Regis, C. Fabio, and B. Boualem, Event correlation for process discovery from web service interaction logs. 2010.

26. Qin, S. and C. Leo, Advances and opportunities in machine learning for process data analytics. *Computers & Chemical Engineering*, 2019. 126: pp. 465–473.

27. Qin, S., Process data analytics in the era of Big Data. 2014

28. Shang, C. and Y. Fengqi, Data analytics and machine learning for smart process manufacturing: Recent advances and perspectives in the Big Data era. 2019.

29. Zhu, Q. and S. Qin, Supervised diagnosis of quality and process monitoring faults with statistical learning methods. 2019.

30. Zhang, Z. and Z. Jinsong, A deep belief network based fault diagnosis model for complex chemical processes. *Computers and Chemical Engineering*, 2017. 107: pp. 395–407.

31. Wu, H. and Z. Jinsong, Deep convolutional neural network model based chemical process fault diagnosis. *Computers and Chemical Engineering*, 2018. 115: pp. 185–197.

32. Liu, D., Y. Xiong, W. Ding, and W. Qinglai, Reinforcement-learning-based robust controller design for continuous-time uncertain nonlinear systems subject to input constraints. *IEEE Transactions in Cybernetics*, 2015. 45: pp. 1372–1385.

33. Lewis, F., V. Draguna, and V. Kyriakos, Reinforcement learning and feedback control: Using natural decision methods to design optimal adaptive controllers. *IEEE Controls Systems Magazine*, 2012. 32: pp. 76–105.

34. Real-time business activity monitoring and analysis of process performance on big-data domains.

35. The integrated effect of BDA, LSS and GM on the environmental performance of manufacturing companies.

36. Li, L., H. Tongtong, and C. Ting, Evaluation on China's forestry resources efficiency based on big data. 2017.

37. Wu, K., et al., Toward sustainability: Using big data to explore the decisive attributes of supply chain risks and uncertainties. 2016.

38. Zhang, Y., et al., A big data driven analytical framework for energy-intensive manufacturing industries. *Journal of Cleaner Production*, 2018. 197: pp. 57–72.

39. A Big Data analytics architecture for cleaner manufacturing and maintenance processes of complex products.

40. Ren, S., Zhang, Y., Liu, Y., Sakao, T., Huisingh, D., & Almeida, C. M. (2019). A comprehensive review of big data analytics throughout product lifecycle to support sustainable smart manufacturing: a framework, challenges and future research directions. *Journal of Cleaner Production*, 210: pp. 1343–1365.

41. Lingitz, L., et al., Lead time prediction using machine learning algorithms: A case study by a semiconductor manufacturer. *Procedia CIRP*, 2018. 72: pp. 1051–1056.

42. Wang, C. and J. Pingyu, Deep neural networks based order completion time prediction by using real-time job shop RFID data. 2017.

43. Teck. Connecting data systems delivers real value at elkview operations. *Connect Magazine*, 2019. 28: p. 12.

44. The Best Day, Every Day. 2020. [2020 accessed on 14/04/2020] available from: https://australianminingreview.com.au/techtalk/the-best-day-every-day/.

45. Sun, X., Z. Hang, T. Fengliang, and Y. Lei, The use of a machine learning method to predict the real-time link travel time of open-pit trucks. *Mathematical Problems in Engineering*, 2018. 2018: pp. 1–14.

46. Soofastaei, A., F. Euler, and Z. Fernando, Haul trucks queuing prediction in open pit mines. *Australian Resources and Investment*, 2020. 14: pp. 29–31.

47. Cox, W., F. Tim, R. Mark, and W. Lyndon, A genetic algorithm for truck dispatching in mining. In *Third Global Conference on Artificial Intelligence*, 18-22 October, 2017, Courtyard Marriott hotel in Coconut Grove, Miami.

48. Ristovski, K., G. Chetan, H. Kunihiko, and T. Hsiu-Khuern, dispatch with confidence: Integration of machine learning, optimization and simulation for open pit mines. 2017.

49. Steyn, J., B. Osvaldo, and B. Gorain, Process analytics: Transforming mineral process plant data into actionable insights. *Mining Engineering*, 2018. 70(9): pp. 18–29.

50. Conger, R., R. Harry, and S. Richard, Inside a mining company's AI transformation. *Mckinsey & Company*, 2020. 8: pp. 12–18.

7

Predictive Maintenance of Mining Machines Applying Advanced Data Analysis

Paulo Martins and Ali Soofastaei

Introduction

Mining is witnessing the Fourth Industrial Revolution, Industry 4.0. The integration of digital and physical mining systems relates to Industry 4.0. Such incorporation allows a large number of data collected by different tools in the various sectors of the mine sites or plants to be processed.

Also, new Industry 4.0 expertise combines products, individuals, equipment, and devices, making it possible to obtain information quickly and in a targeted manner. The vast amount of data gathered by industrial systems includes information concerning mining process processes, incidents, and alarms.

Also, these data can lead to useful knowledge and expertise in the mining process when the data are collected and examined. Advanced data analytics and improved strategic decision-making can be accomplished by interpretive results, offering benefits such as maintenance costs, reducing the time to stop repair work, decreasing replacement component inventory, enhancing repair part life, and raising production speed, safety in operators, and repair checks.

The operational advantages listed are strongly linked to maintenance processes. Maintenance of equipment is an essential parameter in the mining industry that affects operational time and efficiency directly. Machinery and equipment faults must, therefore, be recognized as avoiding production process disturbances.

Different groups of maintenance management strategies exist in the mining industry. The following was described as a simple classification of maintenance procedures in mining:

- Run-to-failure (R2F) or corrective maintenance only occurs after the system or device ceases working. The most straightforward maintenance technique is to stop adding direct costs to the manufacturing and stop and the reconstruction of the parts to be replaced.

- The routine maintenance operation carried out on a schedule or on a process iteration to prevent process/equipment failures is preventive maintenance (PvM) or scheduled maintenance.
- Statistical maintenance (predictive maintenance, PdM) uses statistical methods to decide the need for maintenance. This focuses on the continuous monitoring of the machinery/equipment or process integrity, which only allows maintenance when necessary. It also enables early detection of the failure by using historical data-driven predictive instruments (i.e., the technique of artificial intelligence, AI), honesty variables (i.e., visual features, wear, colors, etc.), methods of statistical inference, and approaches to engineering.

The ultimate maintenance, mandatory maintenance, requires the integration of big data, analysis, machine learning (ML), and AI. PdM takes it a step further through actions to address an inevitable issue instead of merely suggesting acts.

A summary of the maintenance types is shown in Figure 7.1. Different servicing classes have various roles. By choosing the R2F, however, the mining companies delay maintenance measures and take the risk that their assets will be unavailable. At the same time, PvM anticipates maintenance interventions, which lead to a replacement part swap with half-life.

A robust sustaining approach, therefore, must enhance equipment situation, decrease fault rates for the equipment, and minimize maintenance costs while optimizing machinery life. Based on the reasons mentioned above, the PdM approach is among the unique approaches in the era of Industry 4.0, as it can optimize the use and management of mining sites. Some of the advantages of PdM are maximizing equipment use and operation and reducing costs of maintenance and materials and labor.

PdM addresses monitoring for faults or errors before they occur. This maintenance approach can be categorized into three main classifications, i.e., statistical (ST), AI, and model-based approaches, for diagnostic and prognostic purposes and to track equipment specifications.

Preventive Maintenance
Maintenance is performed periodically with a planned schedule

Run to Failure Corrective Maintenance
Fix an equipment, when it breaks

Predictive Maintenance
Continuous monitoring of an equipment using analytical tools

FIGURE 7.1
Mining maintenance types.

In order to approach the model-based way, it is necessary to follow mechanical and equipment theory and to observe ST approaches. In PdM applications, AI approaches were increasingly applied.

ML has emerged as a powerful tool in AI for many applications to develop predictive AI algorithms. ML techniques can handle large-scale data in a complex environment to uncover hidden associations within data. For PdM applications, ML supports robust predictive methods. The efficiency of these applications depends on the ML method being chosen accordingly.

The Digital Transformation

Entry to real-time operational data is an essential factor in achieving effectiveness. The fourth stage of a maintenance strategy is the application of Big Data Analytics in maintenance: predictive servicing. This is referred to as 4.0 maintenance prediction (or PdM 4.0). Such maintenance rates will reduce maintenance preparation times by 20%–50%, increase the uptime of equipment by 10%–20%, and reduce total costs of maintenance by 5%–10%. A four-stage implementation process is now possible to achieve this level of maintenance (Figure 7.2).

Step one – The first step is to create an operational data infrastructure for businesses. For example, the OSIsoft flagship product PI system is a corporate infrastructure that collects and transforms real-time data from sensors, manufacturing equipment, and other devices into rich real-time insights that link sensor-based data to systems and individuals. This first step is essential for subsequent study. Moreover, a single network that improves the reliability of assets will not only allow real-time operational data infrastructure but also boost process efficiency, energy, and water management, climate, health and safety, quality and

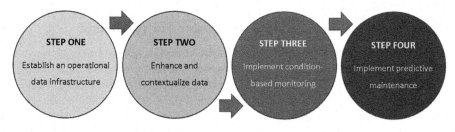

FIGURE 7.2
Four steps of PdM implementation.

Key Performance Indicator (KPI), and reporting. This is the framework for all digital transformation policy projects, such as the PdM 4.0 approach.

Step two – This is how data is saved and updated to information. Improving data means having the data background. Although data is collected from a sensor, analysts must know whether the equipment operates or stops to make the data meaningful. Data have little meaning without context. Also, it is essential to recognize what data is essential and relevant to an organization. The OSIsoft PI System offers businesses contextualized data – the form of data that makes smarter operations possible.

Step three – This concerns the application of condition-based maintenance with contextualized data. This means giving priority to certain assets and defining conditions leading to eventual failure and applying certain conditions on certain assets within an operational data system in real time for automated real-time monitoring. For instance, when a coating temperature increases beyond its average operating temperature, it means that the coating finally fails.

Step four – This is the last level, which is PdM 4.0. In combination with advanced analytics and pattern recognition software, the OSIsoft PI system provides operational intelligence in real time, which enables businesses to optimize their operations.

Both methods together evaluate the trends that lead to a possible failure automatically. The above example allows the question to be asked: how does the temperature of the bearing start to increase in its normal operating range?

This not only increases productivity and lowers maintenance costs when introduced, but mining operations benefit from maximizing their use of resources such as energy and water.

The best way to show how efficient the operational data infrastructure for the OSIsoft PI framework is by using real case studies.

How Can Advanced Analytics Improve Maintenance?

In the digital era, the integration of Big Data Analytics, Cloud Computing, and industrial machinery create opportunities for novel and sophisticated developments in the PdM field. One of the most significant benefits is a substantial reduction in equipment downtime. Advanced PdM can now determine when an equipment component will undergo specific failure, avoiding unexpected events, thus enabling preparatory maintenance to be performed at the right moment. According to [1], significant improvement can be captured through a solid PdM strategy based on advanced analytics. Intelligent systems are a powerful facilitator to allow the optimization of maintenance at low-cost rates. A holistic comprehension of the different equipment, machines, components, and individual parts provides in-depth knowledge of the current status. It enables optimization in the distribution of parts to the right place at the right time. The result is a minimization of inventory materials and maintenance expenses, coupled with machine reliability improvement. Predictive analytics embedded in intelligent systems support engineers to implement preventative maintenance programs that have the potential to enhance machine reliability performance to unprecedented levels [1]. Relevant information and predictive methods lead to strategic improvements in the maintenance, repair, and operations cycle. Predictive analytics can minimize unplanned maintenance by improving fault identification, accelerating root cause finding, accelerating the response to accidents, and optimizing inventory management and assignment.

A recent McKinsey report describes how AI-enhanced predictive analytics can deal with the complexity of forecasting a failure caused by the massive quantity of influencing parameters and the various characteristics of data sources. The capacity of exploring this great diversity of data sources beyond sensor outputs (e.g., maintenance logs, quality measurement of machine outputs, and external data such as weather condition) expands predictions boundaries to new levels. AI-based algorithms can identify errors and differentiate noise from the essential information used to forecast breakdowns and boost future decisions [2]. Based on the report estimation, Figure 7.3 highlights some of the potential benefits of this modern application of advanced analytics for PdM.

Other associated benefits are the reduction of maintenance time, labor, spare parts inventory, safety, health, environment and quality risks, and lifetime maximization.

In order to unlock the value of PdM and achieve business objectives, businesses need to connect their computers, data, insights, and people. Figure 7.4 describes the significant considerations for getting started.

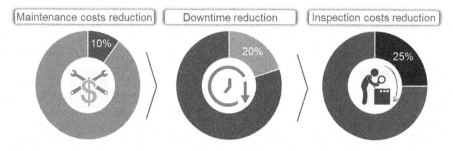

FIGURE 7.3
Benefits of maintenance predictive analytics using AI-based algorithms [2].

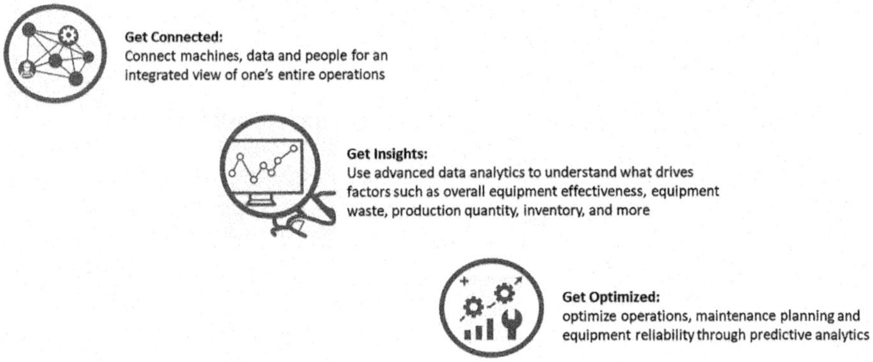

Get Connected:
Connect machines, data and people for an
integrated view of one's entire operations

Get Insights:
Use advanced data analytics to understand what drives
factors such as overall equipment effectiveness, equipment
waste, production quantity, inventory, and more

Get Optimized:
optimize operations, maintenance planning and
equipment reliability through predictive analytics

FIGURE 7.4
Considerations for getting started with PdM [1].

Key PdM – Advanced Analytics Methods in the Mining Industry

The literature indicates that Random Forest (RF) –33%, followed by Artificial Neural Network (ANN), Deep Learning (DL) –27%, support vector machines (SVM), – 25%, k-means – 13%, and others – 2%, are the most commonly used ML algorithms in PdM method [3].

RF Algorithm in PdM

RFs or random decision-making forests are a group learning method for classifying, regressing, and other tasks that operate by constructing a multitude of decision-making trees at training time and generating a class that is the class model (classification) or the mean tree prediction (regression). In 1995, Tin Kam Ho from IBM launched RFs. As the name indicates, the RF produces a "forest" (ensemble) with multiple randomized decision trees and adds a simple average of their predictions. RFs showed good performance when the number of variables exceeded the number of samples (observations). RF is a supervised learning algorithm that is used for tasks of classification and regression.

ANN in PdM

ANNs or connectionist systems are loosely based computation structures that make up animal brains in biological neural networks. Such systems "learn" tasks by taking examples into account, usually without having to program task-specific rules. ANNs are smart, biological neuron–inspired computational techniques. An ANN consists of several processing units (nodes) that work quite simply. In general, such units are connected by the associated weight communication channels and operate with only the local data that is indicated by their connections. The intelligent behavior of ANNs is derived from the interactions between the network processing units. In many industrial applications, including soft sensing and predictive control, ANNs have been proposed and are one of the most common and applied ML algorithms. The main advantages of ANNs are the lack of expert knowledge to make decisions since they are only based on past data; even if the data are inaccurate, they are not compromised and can be used in real time without modifying their design at every update by creating accurate ANNs for a specific app.

Support Vector Machines in PdM

Because of its high precision, SVM is a well-known ML technique for classification and regression tasks. One of the main features of SVM is the high precision with which data classes are separated and also that it is the best way to describe the separating data classes. SVM is a collection of supervised lessons that examines regression and identifies patterns.

k-Means in PdM

The k-means model is common for clusters that use an unmonitored strategy to evaluate multiple clusters. This technique is intended to classify the k partitions/clusters of the datasets so that "close" samples are linked to one cluster, and "far" samples from each other are linked to different clusters. It is easy to apply to the k-means model. This also has good performance and handles massive data sets. When new samples are available, the centers of the clusters can be modified with retraining.

DL in PdM

DL is an ANN in ML, which is characterized by several nonlinear processing layers. The objective is to learn the hierarchical representation of data. It is a fast-growing field with many new types of research and applications being published every week [4]. The next topics will explore the major DL techniques applied in PdM, and some promising applications.

Diagnostic Analytics and Fault Assessment

The latest developments led to improvements in the capacity of aggregate data from smart sensors and autonomous systems. This new landscape enabled the rapid growth of the application of DL techniques in the field of fault diagnosis and classification [4]. By applying Convolutional Neural Network, feature learning and defect diagnosis can be combined in one model and has been used in multiple aspects, such as bearings, gearboxes, wind generators, and rotors [5]. Another promising application involves Deep Belief Network, which has fast inference and presents the advantage of encoding high-order network structures by stacking multiple Restricted Boltzmann Machines. It has been studied for fault diagnosis of aircraft engines, chemical processes, reciprocating compressors, high-speed trains, and wind turbines. Auto Encoder has been tested for unsupervised feature learning; then, the learned features are used as an input in the training and classification phases of traditional ML models. Some potential applications include feature learning from motor current signals and wind turbine fault classifications [5].

Predictive Analytics for Defect Prognosis

PdM solutions developed using Deep Recurrent Neural Network have a strong capacity to model the temporal pattern presented in the historical data found in the manufacturing environment. A general recurrent neural network, known as long short-term memory, has been tested to predict defect propagation and predict remaining useful life of mechanical systems or components [5]. A combination of convolutional neural network (CNN) and bi-directional Long Short Term Memory(LSTM) is used for machine tool wear prediction, where CNN deals with local features extraction from sequential signals, and bidirectional LSTM is responsible for capturing long-term dependency for prediction [6]. Predictive analytics also has repeated to Deep Belief Network as an approach for feature learning in regression models. CNNs with time series imaging has also been proven as an efficient approach to detect wearing condition and location [7].

System Architecture and Maintenance in Mining

The form of maintenance can be modified by general maintenance (e.g., decentralized maintenance) to active preventive significant data maintenance by supporting Wireless communication networks, cloud computing, and Big Data processing. The device architecture for large-scale data production using active PvM is shown in Figure 7.5. There are three levels of maintenance in traditional mode: maintenance of production lines, maintenance of workshops, and maintenance of factories; and deficiencies or issues are not identified in real time from the lower to the upper layer.

Many newer devices may send a warning to a manager to fix new issues as soon as possible following a malfunction or problem. This mode, however, is still not predictable.

An industrial wireless network can collect and forward all related information for active, large-scale PvM, product data, system status, and facility logs; also included are alarms for the system and methods for uploading data processes to the cloud. Correlations may be evaluated in the cloud to determine

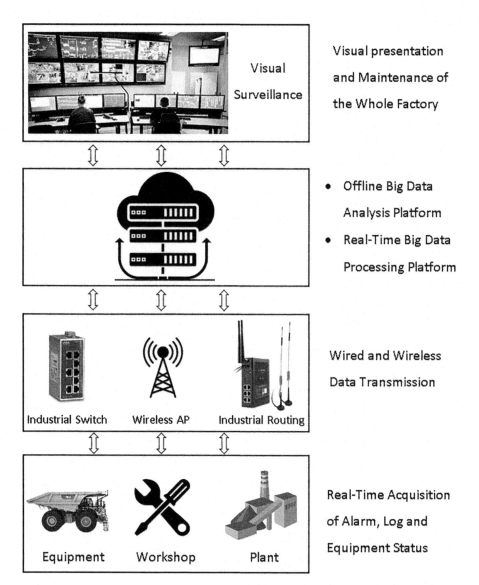

FIGURE 7.5
Big data system architecture for successful PvM.

internal relations. The device status, in conjunction with the facility log, can be used, for example, to discuss the health status of equipment further.

Furthermore, the results can be viewed on a widescreen or via mobile services sent to the manager concerned. The collection system for big data and large data analysis are the main research fields in this architecture. The two components represent the key differences in PvM, both conventional and big data maintenance.

Maintenance Big Data Collection

We primarily focus on cloud data processing in this segment. Wide-scale maintenance data processing is implemented briefly. For significant data-based PvM, all associated data, such as alarms, device logs, and system status, must be collected. The characteristics of these three data types are distinct. Alarms are unpredictable events that should be treated in real time. PC status is obtained regularly, and system logs can be continuously transmitted to the cloud. For ease of transformation and further processing, data formats and pre-processing data are essential for all applications. To do so, OLE for Process Control (OPC) UA was one of the most innovative solutions to data integration, a service-based architecture for industrial applications. The OPC UA has been developed to surpass the capability of the OPC Classics based on the widespread use of the OPC Classic and increased reliability, platform extension, and platform independence aspects. The computer autonomy function enables open communication among different kinds of software and hardware systems with good prospects for a thorough collection of data and superior data analysis.

Network access is an important part of data collection. In general, Zigbee or Wi-Fi is chosen to relay the information collected in functional applications. A Zigbee node requires a coordinator to access the cloud. Therefore, Wi-Fi is more commonly used since no node of the coordinator is needed and Wi-Fi can directly access the cloud. If devices cannot communicate with interface information, an aid to data acquisition must be added. All data collection nodes must follow the Restful protocol to transfer information to the cloud efficiently. Alternatively, data collection nodes also allow the configurable application layer to change sample frequency and protocols.

Framework for PdM Implementation

According to the PWC PdM report, many companies have ambitious plans to implement PdM strategies, but current capabilities are not adequate for full PdM adoption.

Beyond the technological aspect, the scope of implementing PdM is extremely important. Companies need to focus on organizational structure and other relevant factors required to ensure project management and change management skills needed to reach their aspirations [8].

The proposed framework was built to address all these topics, exploring both technological and organizational issues (Figure 7.6).

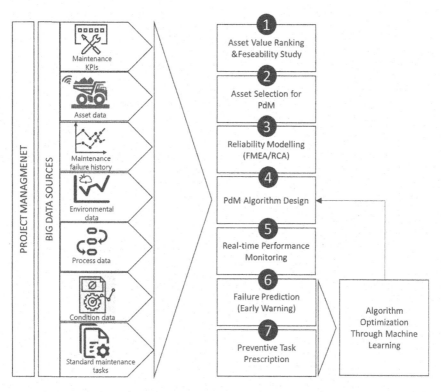

FIGURE 7.6
Framework for PdM implementation [8].

The implementation concept is based on the strategy of gradually deploying the PdM model for some specific assets defined earlier. More details about the seven-step approach are provided below.

1. **Asset value ranking and feasibility study** – Involves the identification of the more relevant assets considering the feasibility and potential for value creation. Moreover, the availability and suitability of data are crucial characteristics to be assessed in advance. The definition of assets is fundamental to develop a solid business case that justifies the required investments.

2. **Asset selection for PdM** – The advice here is to avoid the temptation of covering all assets or plants in one go. Develop pilot projects, and keep the lessons learned from them to apply these to the roll-out of PdM per asset type.

3. **Reliability modelling** – In this step, it is necessary to deep dive into the different failure agents and modes to understand the relation between them and map what type of sensory data, process data, and

external data are necessary. Powerful tools to support these analyses are the root causes analysis and failure mode effects analysis.

4. **PdM algorithm design** – This is the point where the magic occurs. Prediction quality is strictly linked with the type of selected algorithm. The algorithm design task can be significantly simplified when a great model is created in step 3. Sometimes, data scientists are called to build a self-learning algorithm capable of uncovering relevant insights from the pools of data.

5. **Real-time performance monitoring** – After finding a suitable algorithm, it is time to see the PdM model running in a real production environment. This step requires feeding the model with stream data collected directly from the various sources to monitor and visualize the asset performance in real time.

6. **Failure prediction (early warning)** – Taking actions based on algorithms outputs implies a mindset change. Initially, reluctant behavior is expected, especially if management and front-line employees are not familiar with data analytics. General good practice, in this case, is to run the PdM model in parallel with current maintenance procedures until everyone has confidence in the predictions.

7. **Preventive task prescription** – This is the highest level of intelligence provided by a PdM model. The failure prediction capability is complemented by the algorithm's ability to prescribe the best action to avoid failure. A set of standard maintenance tasks is used to train the model. Some more advanced implementations include the automatic execution of related work orders.

The implementation of PdM is not only a technological challenge. Project management skills are also required to deliver the expected outcomes successfully. Moreover, the continuity of the results reached by PdM efforts is also linked to organizational alignment. In order to consecutively perform all the seven steps described above, a solid Big Data structure is essential. The next section will explore the requirements for PdM in this Big Data era.

Requirements for PdM

The introduction of digital technologies in the mining industry has two main objectives: to maximize the uptime throughout the production chain and improve productivity while reducing production costs. Nowadays, the exciting journey towards these objectives is surrounded by Big Data and IoT technologies, and PdM plays a central role. Organizations need to understand the prerequisites for processing big data in different phases of the analytical cycle, including data collection, analytics, querying, and storage [9].

In recent research published by [9] and [2] use case requirements, and associated subitems were described considering queuing management, platform, storage, and SQL engines. Figure 7.7 summarizes the related requirements for PdM.

Distributed queuing management collects equipment sensor data. In order to perform this task, the system needs to be efficient and scalable to quickly and accurately gather data from several machines. After collecting the available data, it is necessary to represent it according to its characteristics (structured, semi-structured, or unstructured) and determine a better storage model (distributed file system, document-based, column-based, graph-based). Additionally, considering the CAP theorem, the selected big data storage solution needs to have an efficient retrieval capability to feed the PdM solution in a timely and correct manner. Typically, the natural choice is an in-memory data processing solution, which provides fast data processing suited for connecting with big historical data or streaming data. On the flip side, batch and stream processing can also be developed using big data stack technology. The final definition of the big data platform depends on the PdM application [9].

The next requirement takes into consideration the different types of data and their intrinsic complexity. Data collected may be in multiple formats

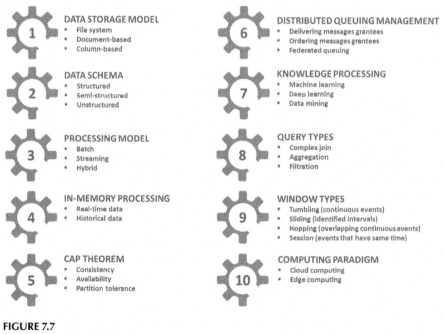

FIGURE 7.7
Big data requirements for PdM [9].

such as text, videos, images, audio, process signals, graphics, and time-series sequence data. Therefore, the knowledge processing approach selected typically involves ML, data mining, or DL methods to predict failures accurately and enable smart decisions. Furthermore, the work of processing the massive and multivariate time-series-generated data requires window-based queries such as join and aggregation. The oncoming event characteristics determine the type of window. The final requirement described in the study is regarding the deployment of the IoT-based PdM solution, which considers the computing trade-off to perform analysis—for example, cloud computing or edge computing [9].

Cases and Applications

In the survey conducted by [10] on some mining companies in Australia, the conclusion was that PdM, Big Data, IoT, and Data Analytics create safe work environments, enable better maintenance practices, and reduce maintenance costs. More than one-third of the total mining costs are destined for maintenance processes, which makes it the highest controllable cost. The majority of failures do not occur instantaneously, and normally there are few signs of degradation processes or trends that indicate the transition from normal status to failure [11]. Supported by the emerging technologies, miners can collect and assess valuable data and make more reasonable maintenance decisions.

Digital Twin for Intelligent Maintenance

The smartest level of advanced analytics for maintenance can be achieved through the implementation of Digital Twins (DT). In this advanced approach, all information collected about an individual physical asset or combination of assets in a system is used to maximize asset utilization and optimize its performance. The historical data used to model a DT includes operational conditions under which it has been used, its configuration, maintenance events, and other exogenous conditions, among others. The DT is designed to enhance human effectiveness, productivity, and capacity by focusing on more strategic business issues [1].

In mining applications, the models start by guiding "design limits" and, then, are continuously updated and learn to efficiently mirror the asset under different operational scenarios and related variations – ore types, temperature, weather conditions, air quality, moisture, load, weather forecast models, and more. Combined with advanced technologies, DT models can reach the next level of optimization, control, and prediction, delivering more accurate results for asset performance, reliability, and maintenance. Additionally, by using the sensor data, models enable mitigation of threats and unplanned downtime with the evaluation of different scenarios, understand trade-offs, improve efficiency, and ultimately perform PdM. DT solutions are the future of mining asset management, and the number of successful implementation cases are constantly growing. Figure 7.8 depicts the use cases of DT models across the entire asset's lifetime, starting with the design phase.

According to Kevin Shikoluk, a global strategic marketing leader for Digital Mine at GE, DT goes beyond the prediction that an equipment or component will breakdown. DT models show what will happen if the problem is not fixed. They allow its users to test different changes in real time before executing them in the plant [12,13]. Some mining giants like Rio Tinto, Anglo American, BHP, and Newmont are embracing DT solutions not only for individual assets but to digitalize the entire operation [14–16].

PdM for Mineral Processing Plants

Mineral processing plants are equipped with a high number of fixed assets that perform essential transformations in the ore feed. Equipment downtime directly results in production losses and high costs. Although most of the processing units are well instrumented and connected, only a few miners are effectively using this data to improve their maintenance performance.

In 2018, the Barrick Gold Cortez Mine, in Nevada, deployed a PdM project in the gold refining process applying ML and equipment sensors in order to identify potential issues before they escalated into failure. The algorithm generates a healthy asset score using collected sensor data and tracks relevant

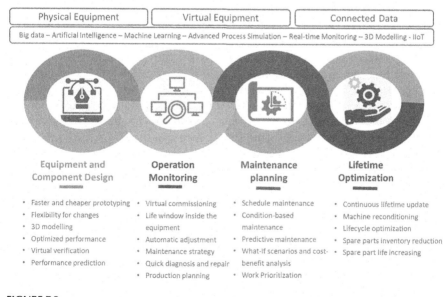

FIGURE 7.8
DT capabilities in all phases of asset life.

variations that might indicate undesired failures. After implementation, the investment paid off in a short period since several major failures were detected and avoided. For example, one of the fault predictions alone saved $600,000 [17,18].

At Newcrest Mining's Lihir operation in Papua New Guinea, ML algorithms were applied to reduce unplanned breakdowns in its semi-autogenous grinding mills. The project was developed in partnership with the technological solutions company Petra Data Science, which also provides digital twin solutions for mining operations. Taking data going back one year, the team focused on the critical signals or indicators associated with overload events. This big data included noise, power, speed, energy consumption, and control variables, most of them recorded at five-second intervals. The model was able to find patterns and anomalies in the data and use them to predict the probability of an overload event one hour in advance, enabling operators and engineers to proactively take the required actions to avoid it [19].

In a partnership between the industrial AI software company Uptake and the world's largest copper producer, Codelco, an AI solution is implemented to monitor the health processing plant (and mining) equipment through an enterprise-wide Asset Performance Management solution across all of the mining sites. The project's approach will generate industrial data science original equipment manufacturer-agnostic insights, predictions, and prescriptions for any asset [20,15].

Belt transport systems play an important role in the mine's ore transport system. In order to meet operations requirements, an online fault detection system is essential. In [21], multivariate analytical models based on data fusion and artificial intelligence techniques were implemented to avoid belt conveyor failures, supporting planned repairs, and reduce repair costs and production losses associated with breakdowns.

In [15], a data science team developed an anomaly detection algorithm that was able to determine the indicators of failure in a Ball Mill drive gearbox and pinion bearings. Although there was a large amount of condition monitoring data available (vibration, pressure, and temperature sensors), only a small number of failure records were available. The challenge was to isolate the different failure mechanisms (no failure classifications provided) and develop a predictive model that would allow component repair/replacement to be scheduled during quarterly plant shutdowns. The developed model was tested against previous failure data to determine the appropriate confidence level of the model. Back testing results show 85% accuracy in predictions, correctly identifying past failure events separately to repair events.

PdM for Mining Fleet

A mining fleet is the most critical and expensive asset of a mining operation. Operating and maintaining that equipment at high performance is expensive, and unplanned maintenance events have a direct impact on the production outcome and costs. The failure of a single component could affect the entire system, and one day of unexpected downtime could cost many thousands of dollars [22]. Mining operations worldwide typically adopt preventative maintenance based on the equipment's manufacturer schedules. These estimations can be drastically affected by the actual use of the equipment.

According to the literature review described in [22], multiple approaches can be used to perform predictive analytics in a fleet.

- **Summary statistics** – Evaluates the behavior of equipment to learn relevant statistics. For example, counting faults and determining equipment efficiency.
- **Hypothesis testing** – Comprises the investigation of causal relationships.
- **Clustering** – Applied to find novel concepts and allows heterogeneous groups of items to be grouped by similarity. For instance, fault categories can be grouped according to sensor data.
- **Classification** – Can be used to categorize a variety of faults.
- **Anomaly detection** – Can be applied to find abnormal signals due to a fault or malfunction.
- **Frequent pattern mining** – Can identify correlations between variables and could be used to search faults that often occur together.
- **Process mining** – Used to reconstruct sequences of activities using event data and may be applied for deviation detection.
- **Sensor selection** – Approach used to improve the accuracy of the chosen models.

The analytical evaluation of the fleet can be categorized into two major types: onboard analytics and fleet-wide analytics. In the onboard analytics, data is processed locally in the equipment, which means that only a single equipment data is used in the predictive models. The benefits of this approach include (1) fog computing (a computational resource that enables real-time analytics and supports low latency and lightweight computing) and (2) the capacity of supporting simple models to analyze individual vehicle's sensors, such as expert systems. On the other hand, fleet-wide analytics combine data from the entire fleet, performing cross-fit analytics. The benefits include (1) higher accuracy in identifying faults, (2) capacity of accurately identifying no-fault events, (3) uncovering new types of faults, by applying similarity approach, (4) capacity of searching for root causes of the faults and (5) great performance for monitoring the normal status of many similar vehicles (a most common scenario) since fault data is often scarce. Finally, it is known that systems that integrate both fleet-wide and onboard analytics produce better PdM results [22].

In terms of advanced analytics techniques for PdM, a variety of approaches are found in the literature. In underground mining, predictive failure using data from Load Haul Dump (LHD) is a rich research field. A typical LHD has more than 150 sensors.

Using data from an underground mine in Canada, [23] studied the application of genetic algorithms for PdM of mining machinery. Assuming that failures of mining equipment caused by various agents follow the biological evolution theory, the team created software that analyzed data of an LHD and reached satisfactory results. Other successful applications involve supervised and unsupervised ML algorithms. According to Ref. [24], a supervised learning approach using Support Vector Machine and NN can be used to classify engine condition of an LHD in three categories: normal, suboptimal, or imminent failure. The model's inputs considered for training the algorithm included engine data (torque, coolant temperature, oil pressure, turbo boost pressure, and speed), wheel-based speed, and historical based speed. Another ML case relied on unsupervised ML using One-Class Support Vector Machine to predict conditions leading to engine failure.

In the thesis published in Ref. [11], the Sequential Pattern Mining technique was used for analyzing big data collected from a North America mining company. The dataset presents records from eleven trucks for nine months. Sequential Pattern Mining is a data mining technique that searches for patterns that occur consecutively in a database or patterns that have an association with time or other values. The approach has demonstrated great applicability to uncover important patterns in sequential data. In the studied case, three failure codes were selected and, first, several patterns between the two same codes were identified. Second, a variety of patterns were uncovered in the last three and five shifts that anticipate the breakdown. Despite some improvement opportunities reported by the author, the prediction rate was more than 90% in the last five shift events.

The mining company Barrick Gold decided to improve the asset health monitoring system of their truck haul fleet in order to increase maintenance efficiency and reduce costs. The operation where the initiative was developed – Pueblo Viejo – had 34 haul trucks, and the project involved real-time gathering information using the available systems and sensors at minimum cost. Instead of keeping with the traditional reactive approach, the maintenance team is now one step ahead of the failure. As a result, the company saved $500,000 and reduced the total number of failures from engine, brake, and suspension faults by 30% [17].

In a pilot study with a leading mining operator, the industrial AI software company Uptake processed 365 million data points from a fleet of 44 Caterpillar 797Fs to uncover insights that resulted in a 3% downtime reduction and a 5% improvement in haul truck utilization. The most difficult challenge reported was related to adding context to the machine fault codes before training the ML model so that the false failures could be distinguished from true failures that require action [24].

A predictive analytics solution provided by the PdM software company Dingo provides predictive models based on AI and ML techniques. The Anomaly

Detection and Remaining Useful Life models are developed by collecting and combining failure data from the actual equipment. One of Dingo's mining clients was experiencing a degrading final drive life on Caterpillar 789 C&D trucks, which resulted in significant costs and downtime impacts to the operation. Before the implementation of the predictive analytics solution, the average life of final drives had decreased by more than 30% (from 19,092 to 13,229 hours). After understanding the most common failure modes (broken/worn teeth in the central gear and worn copper washers) and using the only condition monitoring data available for these components (oil analysis), a team of data scientists identified the most correlated oil analysis indicators to failure and developed proprietary ML models to predict future failures.

References

1. Daily, J. and P. Jeff, Predictive maintenance: How big data analysis can improve maintenance. In Richter, K., Walther, J. (eds.) *Supply Chain Integration Challenges in Commercial Aerospace.* 2017: Springer, Cham.
2. Mckinsey. Smartening up with Artificial Intelligence (AI). 2017.
3. Carvalho, T., et al., A systematic literature review of machine learning methods applied to predictive maintenance. *Computers & Industrial Engineering,* 2019. 137: p. 106024.
4. Zhao, R., Y. Ruqiang, C. Zhenghua, W. Peng, and G. Robert, Deep learning and its applications to machine health monitoring: A survey. 2015.
5. Wang, J., M. Yulin, Z. Laibin, G. Robert, and W. Dazhong, Deep learning for smart manufacturing: Methods and applications. *Journal of Manufacturing Systems,* 2018. 48: pp. 144–156.
6. Zhao, R., Y. Ruqiang, W. Jinjiang, and M. Kezhi, Learning to Monitor Machine Health with convolutional bi-directional LSTM networks. *Sensors,* 2017. 17: p. 273.
7. Martinez-Arellano, G., T. German, and R. Svetan, Tool wear classification using time series imaging and deep learning. *The International Journal of Advanced Manufacturing Technology,* 2019. 104: pp. 3647–3662.
8. PwC and Mainnovation. Predictive Maintenance 4.0: Predict the unpredictable. 2017.
9. Sahal, R., B. John, and A. Muhammad, Big data and stream processing platforms for Industry 4.0 requirements mapping for a predictive maintenance use case. *Journal of Manufacturing Systems,* 2020. 54: 138–151.
10. Fekete, J., Big data in mining operations. *Copenhagen Business School,* 2015.
11. Kahraman, A., Maintainability analysis of mining trucks with data analytics. 2018.
12. Kruczek, P., et al., Predictive maintenance of mining machines using advanced data analysis system based on the cloud technology. 2019. 5, pp. 34-37.
13. Moore, E., Digital double. *CIM Magazine,* 2018, 2: pp. 46–51.
14. Kirkwood, B. and D. Emilie, Artificial intelligence in mining: It has begun. *IDC Perspective,* 2020.

15. Dingo. The practical application of predictive analytics. *Austmine Webinar: Operating with a Crystal Ball*, 2020.
16. Gleeson, D., Codelco Looks to Uptake's AI Solution for Equipment Maintenance Gains. 2020. [accessed 29/04/2020]; available from: https://im-mining.com/2019/03/26/codelco-looks-uptakes-ai-solution-equipment-maintenance-gains/.
17. Provencher, M., A Guide to Predictive Maintenance for the Smart Mine. 2020 [accessed 29/04/2020]; available from: https://www.mining.com/a-guide-to-predictive-maintenance-for-the-smart-mine/.
18. Rana, T., Z. Leonard, and K. Abhinav, Optimizing predictive maintenance at barrick gold. *PI World Barcelona*, 2018.
19. Leonida, C., Algorithms: Mining's crystal ball. *Mining Magazine*, 2017. 6: 18.
20. Uptake. Improve throughput and reduce downtime in processing operations. 2019.
21. Stefaniak, P., W. Jacek, and Z. Radoslaw, maintenance management of mining belt conveyor system based on data fusion and advanced analytics. 2018.
22. Killeen, P., Knowledge-based predictive maintenance for fleet management. 2020.
23. Peng, S. and V. Nick, Maintainability analysis of underground mining equipment using genetic algorithms: Case studies with an LHD vehicle. *Journal of Mining*,. 2014: pp. 1–10.
24. Uptake. Drive more value from your fleet of haul trucks. 2019.

8

Data Analytics for Energy Efficiency and Gas Emission Reduction

Ali Soofastaei

Introduction

Since the rise in fuel costs in the 1970s, the value of energy consumption has gradually increased. Given that petroleum products are the main sources for energy used in the mining industry, such as power, coal, and natural gas, raising margins can also minimize millions of tons of gas emissions.

In our national security, the national economy, and the lives of all, mining plays a vital role. To maintain living standards, millions of tons of materials should be mined every year [1]. Mining is an essential part of the global economy. It offers vital raw materials such as steel, metals, minerals, sands, and gravel to production and construction sectors, utilities, and other enterprises, [2]. In other words, for many years, mining will continue to be an important part of the global economy.

Mining is an intensive task in terms of energy. In the USA, mining is, for example, one of the few non-produced industries classified as energy-intensive by the US Department of Energy [3]. It is also well known that the mining sector can increase its energy efficiency significantly. Again, the US Department of Energy estimated that the US mining sector uses approximately 1315 PJ per year. This annual energy consumption could be reduced to 610 PJ or about 46% of the current annual energy consumption [3]. Based on recent figures, the Australian mining sector had an energy consumption of about 730 PJ in 2017–2018, up 9% from the previous year. This is somewhat higher than the average energy consumption rises in the last decade. Mining in South Africa consumes 175 PJ of energy a year and is the largest electricity user at 110.9 PJ a year. The effect of this energy intensity on mining operating costs is evident in the connection between increased energy efficiency interests and energy prices [4]. Because of recent policy interventions by various governments to implement industrial costs associated with carbon emissions (carbon taxes and related regulatory costs), these high-energy operations are not appropriate from a sustainable or cost point of view. Therefore, all stakeholders have a deep interest in improving mining energy efficiency.

Mining companies are both considering reducing energy use to lower costs and reducing emissions, especially because of the implementation of carbon emissions schemes. To do this, businesses need to have a clear understanding of their current use of resources.

Mining companies are engaged in the review of their finances, capital costs, and operating plans to ensure that their activities will be profitable and environmentally friendly. Tangible savings must be obtained by investments in sustainable practices and capital equipment. In order to reduce both costs and environmental effects, mining companies aim to improve energy efficiency.

Sustainable investments have not seen substantial returns on investments in previous years. However, returns are increasingly attractive due to the rapid shifts in the regulatory and economic environment. Investments are much more desirable when the advantages of modern technology and business practices are taken into consideration, including direct savings from increased efficiency and opportunities like carbon tax credits. For example, these same investments in energy savings are also extremely desirable when you look at them in the long term. The energy (and financial) savings incentive has stimulated mining and government work into reducing energy consumption. To this end, many research studies and industrial enterprises have been undertaken in mining operations worldwide. Full implementation of state-of-the-art technology and deployment of new technologies through investments in research and development would save about 37% of mining's current energy consumption.

The deeper depth of the mining ore, which requires greater production, transport, and process efforts, increases mining resources. Mining operations use energy in various ways: for excavation, material transfer, milling and processing, ventilation, dewatering, etc. Thus, there are substantial incentives to reduce energy consumption based on the experience of completed industrial projects. All governments and the mining industry have been inspired by the opportunity to curb their energy use.

Data analytics is a very suitable approach in which the different data sources are collected because it is a science to research raw data in order to form conclusions. Cost reduction, faster and better decision-making, and eventually, new products and services can be the major advantages of data analysis [5].

The four main phases of the mining process that data analytics can be used in are the (1) extraction of ore, (2) materials handling, (3) ore comminution and separation, and (4) mineral processing. The focus of many companies is efficiency improvement in the materials handling phase. The transport operation at an open-pit mine constitutes a significant share of total energy consumption. Improvement of the energy efficiency of trucks would, therefore, reduce fuel consumption (FC) and emissions of greenhouse gases (GHGs).

The truck and shovel operation in surface mining is the most common method of extraction and transport of materials [6,7]. Transport overloads form a significant part of energy usage [8]. Such parameters depend on how much energy is consumed. Carmichael et al. conducted a study on the effects of load density, site geology, road surfaces, and gradients on transport trucks 'energy consumption [9]. Cetin analyzed the links between transport operators' energy efficiency and load rates, vehicle performance, and driving habits [10]. Beatty and Arthur studied the effect of certain general variables on energy consumed by trucks, including mine preparation and cycle time [6]. To minimize FC in transit operations, the optimum values of these parameters are calculated. The Coyle study focuses on the effect of payload on truck fuel use. The effect of the load density variability on the FC of freight vehicles is shown in this analysis [11]. Soofastaei et al. have completed several different energy efficiency projects in surface and underground mining haulage trucks [5,12–19].

The literature was mainly based on the statistical methodologies used to estimate the consumption of fuel by mine trucks, to the best of the knowledge of the authors. These models are based on the curves designed for the performance of cargo mining trucks by the truck manufacturer [7,20–25].

Data analysis provides a very appropriate way of pooling these different sources of data since raw data is analyzed in order to come to a conclusion. The most benefits of data analytics can be presented by cost reduction, faster and better decision-making, and finally, new products and services [5]. Data analytics is widely used and can be used in areas that many might not have thought about before. One area that sees much potential in data analytics is the mining industry. For an industry that does trillions of dollars in business every year, data analytics should be considered a necessity, not a luxury.

One of the advanced data analytics techniques in this chapter tackles the important issue of energy efficiency in mining. The aim will be on haulage activities at the open-pit mines. This work is designed to develop an advanced data analysis model to examine the complex interactions that affect transport trucks' energy efficiency in surface mining. The goal of this study is to use the Artificial Neural Network (ANN) to simulate predictions and the Genetic Algorithm (GA) to optimize energy efficiency analysis. This study examined the results of the three key effective FC parameters of haulage trucks. These are payload (P), Speed truck (S), and Total resistance (TR) parameters. Assessing the connection between FC and the above parameters on a real mine site is difficult. Thus, two artificial intelligence approaches were used in this project to create a model for estimating and reducing FC. Thus, two artificial intelligence approaches were used in this project to produce a prototype for estimating and reducing FC.

Advanced Analytics to Improve the Mining Energy Efficiency

Mining Industry Energy Consumption

Global mining companies operate under economic and regulatory constraints, which are currently very difficult. Many organizations in the sector report their success now in this field in response to increased social concerns about the various impacts of the minerals sector and the advent of the idea of sustainable development. However, the use of energy and its climate change impact are priorities, with sustainable companies reporting the overall energy use and the associated GHG emissions.

Mining companies set targets to improve these indicators. However, at the same time, there are global trends towards more complex and less energy-intensive orebodies. In order to increase the environmental sustainability and productivity of their activities, mining companies must, therefore, be more creative. In order to reduce their GHG emissions, companies must, in particular, take into consideration the specific energy usage of their operations.

Research by the government of Australia revealed that the most energy-consuming industries in 2018–2019 were the transportation product, oil, gas, and mining industries. Australia uses a quarter of its annual transport capital. The manufacturer of metal goods such as aluminum, steel, nickel, lead and iron, zinc, copper, silver, and gold accounts for almost 16% of energy consumption. The mining industry absorbs 10% of the energy consumption of participants. Figure 8.1 displays the other industries with the most significant energy use in 2018–2019.

Milling (40%) and material handling by diesel equipment (17%) is the primary type of energy-consuming machinery in the mining industry [16].

Data Science in Mining Industry

Data analysis is a technique used to analyze raw data to find useful information, to come to conclusions about the importance of the data and to promote decisions. The critical incentive presented for the mining of data analytics is the ability of the organization to define, recognize, and then direct the correction of complex root causes, which entail high costs. Data analysis can, therefore, reduce costs and speed up better decision-making, allowing the production and implementation of new products and services and generating added value for everyone [5].

Figure 8.2 illustrates the two dimensions of maturity: a time dimension (over which capability and insights are developed) and a competitive advantage dimension (the value of insights generated). At the lowest levels, analytics are routinely used to produce reports and alerts. These use simple, retrospective processing and reporting tools, such as pie graphs, top-ten histograms, and trending plots. They typically answer the fundamental

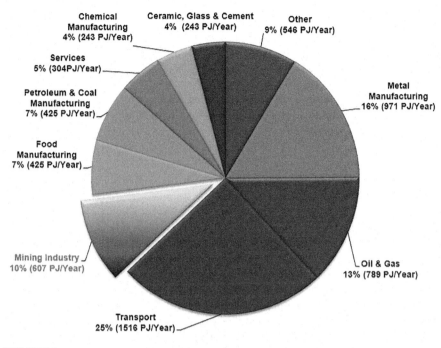

FIGURE 8.1
Top energy users by industry sector (Australia, 2018–2019).

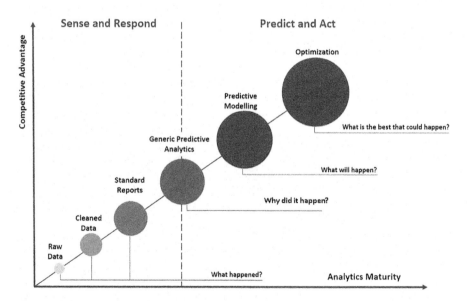

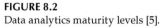

FIGURE 8.2
Data analytics maturity levels [5].

question: "what happened and why?" Increasingly, sophisticated analytical tools, capable of working at or near real time and providing rapid insights for process improvement can show the user "what just happened" and assist them in understanding "why" as well as the next best action to take. Towards the top end of the comparative advantage scale are predictive models, and ultimately optimization tools, with the capability to evaluate "what will happen" and the ability to identify the best available responses – "what is the best that could happen?"[5]

Haul Truck FC Estimate

Mine trucks consume fuel depending on some parameters (Figure 8.3). The seven main groups of these criteria include fleet management, mining preparation, advanced machinery, road haulage, construct and production, environment, and energy efficiency [12,24].

This chapter addressed the impact of P, S, and TR on the energy consumption of mine vehicles. The total resistance equals the total resistance to the grade (GR) and the rolling resistance (RR) [24].

$$TR = GR + RR \tag{8.1}$$

The RR is dependent on the tire and road surface. The Rimpull Force (RF) is the force that prevents motion while the truck pneumatic rolls on the road. The standard value range for RR is between 1.5% and 4.0%. However, the

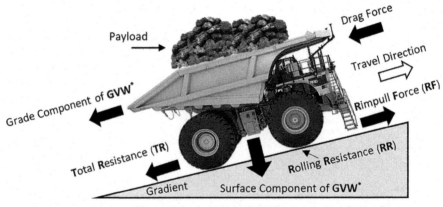

*Gross Vehicle Weight (GVW) = Load + Truck Weight

FIGURE 8.3
Road and truck relevant parameters [14].

RR in the mud with a soft spongy foundation can be more than 10% for road conditions [12].

The GR is the gradient of the road and is proportionally expressed and taken as a ratio between the horizontal and long ascent[12,26]. For instance, a segment of the road up to 15 m over 100 m has a GR of 15%. The GR may be positive or negative depending on a truck moving up or down the ramp. The correlation of the above parameters is demonstrated by the Rimpull-Speed-Grade technical capacity of the truck manufacturer (Figure 8.4).

The mine haul truck FC from Equation (8.2) can be estimated [26]:

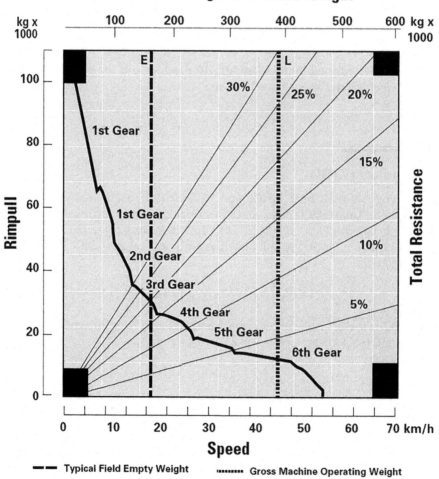

FIGURE 8.4
Rimpull-speed-grade haul truck curve (CAT 793D) [25].

$$FC = (SFC \times LF \times P_o)\%FD \tag{8.2}$$

where SFC is a full-power Particular Fuel burned (0.213–0.268 kg/(kw.h)) and FD is the density of fuel (0.85 kg / L Diesel). Equation (8.3) shows Equation (8.2)'s simplified version [27].

$$FC = 0.3(LF \times P_o) \tag{8.3}$$

where LF is the load factor of the engine and is calculated to the average payload percentage for an operating period [28]. The standard LF values are shown in Table 8.1 [24]. P_o is the truck power (kW) in Equation (8.3) and is defined by

$$P_o = (RF \times S)\%3.6 \tag{8.4}$$

The RF is determined by the Rimpull (R) product and the gravity acceleration (g) and S is the Truck Speed.

Emissions of GHG

Diesel engines emit both GHG_S and Non-Greenhouse Gases [29] into the atmosphere. Total GHG emissions are estimated corresponding to the Global Warming Potential and stated in CO_2 equivalent or CO_2-e [30,31]. The following equation is used to calculate the emissions of the GHG_S diesel truck engine [30,32].

$$GHG_{Emissions} = (CO_2 - e) = FC \times EF \tag{8.5}$$

where FC is the quantity of fuel consumed (kL) and EF is the emission factor. EF for haul truck diesel engines is $2.7\, t\, CO_2 - e / kL$.

TABLE 8.1

Regular Load Factor Values(LF) [24]

Conditions of Service	LF (%)	State of the Road
Low	20–30	Uninterrupted service with a moderate gross weight of the vehicle less than proposed. The mission is not over.
Average	30–40	Constant running on standard gross vehicle weight, minimum payload suggestion
High	40–50	Continuous service with or above the estimated cumulative gross vehicle weight

Mine Truck FC Calculation

This study explains the relationship between combusted truck fuel and established factors (P, S, and TR). The subsequent section describes an ANN model developed to assess the variability in the truck's FC.

Artificial Neural Network

ANNs are a growing synthetic intelligence framework to model multiple variables with one essential factor fitness. Considering specific parameters that influence the FC of mine vehicles, FC can be determined. ANNs are applied in various engineering fields, such as raw material engineering [33–35], chemical engineering [36], and motorized engineering [37–39]. ANNs are needed to respond to multifaceted challenges since they can understand the combined relationships between the various problems involved. ANN's main benefit is that it allows nonlinear and linear correlations to be simulated between factors by using the data provided for learning the network. ANN is the representation of templates that are used for learning by the brain, also known as a parallel distribution [39]. They are a variety of techniques that mimic some of the established features of standard nerve structures and rely on analogies of adaptive learning that have been accepted. The critical part of an ANN model may be the unique structure of the information processing classification. Weighted connectors, therefore, provide enough neuronal mapping. ANNs are used in several computer applications for multifaceted issues.

An ANN was developed in this chapter to produce a FC index (FC_{Index}) as a function of P, S, and TR. The defined parameter indicates the number of liters of diesel consumed in an hour to push one ton of mined material.

Modeling Built

A feedforward, multi-layer perceptron NN with three input variables and one output is the configuration of the generated ANN algorithm for function estimates. The functions of activation in the hidden layer (f) are the continuous nonlinear sigmoid tangents given in Equation (8.6).

$$f = \tan \text{sig}(E) = \frac{2}{1 + \exp(-2E)} - 1 \tag{8.6}$$

where E by Equation (8.7) can be calculated.

$$E_k = \sum_{j=1}^{q} (W_{ijk} X_j + b_{ik}) \quad K = 1, 2, , \ldots .m \tag{8.7}$$

In cases where x is the normalized input differ, w shall be the weight, i shall be the input, b shall be the bias, I shall be the number of input variable, and k and m shall be the number and number of neural network nodes in the hidden layer, respectively.

Equation (8.7) can be used to activate hidden layers and output layers (in this equation, F is the transfer function).

$$F_k = f(E_k) \tag{8.8}$$

The manufacture cap calculates the weighted sum of the signals and related coefficients given by the hidden layer. Equation (8.9) allows for network efficiency (Figure 8.5).

$$\text{Out} = \left(\sum_{k-1}^{m} W_{ok} F_k \right) + b_o \tag{8.9}$$

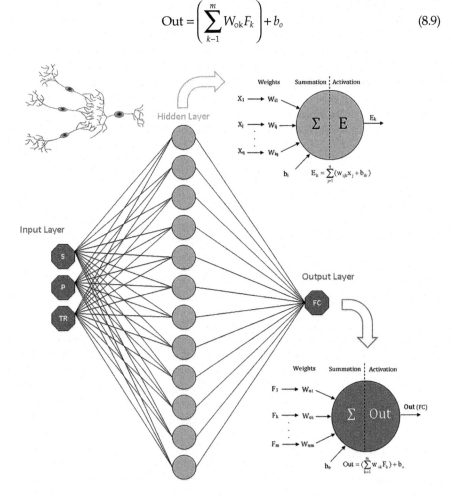

FIGURE 8.5
The scheme structure of the ANN (Example).

Application Established Network

The developed ANN algorithm can be used for estimating truck FC, according to P, S, and TR, based on the steps below.

Step 1 – Standardization of input parameters −1 to +1

$$X_n = \left(\frac{X - X_{min}}{X_{max} - X_{min}} \times 2 \right) - 1 \tag{8.10}$$

Step 2 – The E parameter calculation for each hidden node

$$E_k = \sum_{j-1}^{q} \left(W_{i,j,k} X_j + b_{i,k} \right) \quad k = 1, 2, \ldots, 15 \tag{8.11}$$

Step 3 – F parameters estimation

$$F_k = \frac{2}{1 + \exp(-2E_k)} - 1 \quad k = 1, 2, \ldots, 15 \tag{8.12}$$

Step 4 – Normalized FC Index Estimate (FC$_{Index(n)}$)

$$FC_{index(n)} = \sum_{k-1}^{15} \left(W_{o,k} F_k \right) + b_o \tag{8.13}$$

Step 5 – FC$_{Index(n)}$ denormalization

$$FC_{index} = 13.61 + \frac{(FC_n + 1)(237.92 - FC_n)}{2} \tag{8.14}$$

Applied Model (Case Studies)

The evaluation and validation process of the developed model was based on several data sets from mine engineers obtained in large surface mines in Australia and the USA. Table 8.2 includes complete detail on these mines.

In order to train the established ANN model, only a small number of pairing data were randomly collected from mines (Table 8.3). Independent samples were used to check the network accuracy and validate the model. The results show that the real and projected FC values at all the investigated mine

TABLE 8.2

Studied Mine sites (General Information)

Name	Product	Location	Fleet Size
Blackwater	Coal	Queensland, Australia	184 Truck
Morenci	Copper	Arizona, USA	165 Truck
Sierrita	Copper	Arizona, USA	76 Truck
Kayenta	Coal	Arizona, USA	83 Truck

TABLE 8.3

Generated Model Training and Validation Data Sets

Mine Site	Used Data for Training	Used Data for Validation
Blackwater (Australia – Queensland)	1,500,000	2,000,000
Morenci (USA – Arizona)	1,000,000	1,500,000
Sierrita (USA – Arizona)	2,500,000	3,000,000
Kayenta (USA – Arizona)	800,000	1,000,000

sites are appropriate. In the test results of the synthesized networks, the horizontal and vertical axes are shown in Figures 8.6, where the approximate FC values are shown by the actual FC and the model.

Product Results Established

Figures 8.7–8.10 show the relationship between the P, S, TR, and FC_{Index} provided by the established ANN for a standard range of payloads in four studied mining sites for different truck types. The graphs indicate that FC_{Index} and Gross Vehicle Weight (GVW) have a nonlinear relation. The empty truck weight plus payload is GVW. The rate of energy consumption increases intensively with rising total resistance. However, by adjusting truck speed, this energy consumption rate does not change suddenly. The model built also shows that the sum of FC_{Index} varies by truck speed

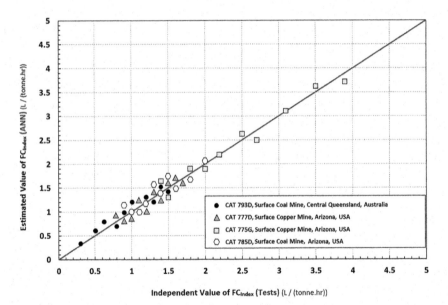

FIGURE 8.6

Comparison of the actual values with the expected FC for trucks by the established ANN model.

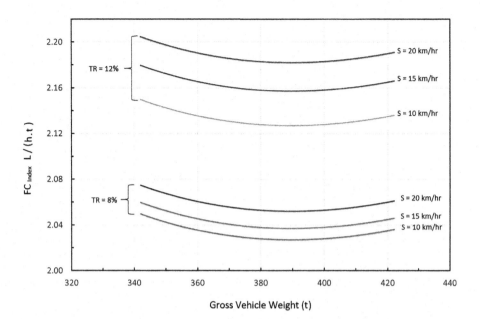

FIGURE 8.7
Correlation calculated by ANN between *P*, *S*, TR, and FC$_{Index}$ (CAT 793D).

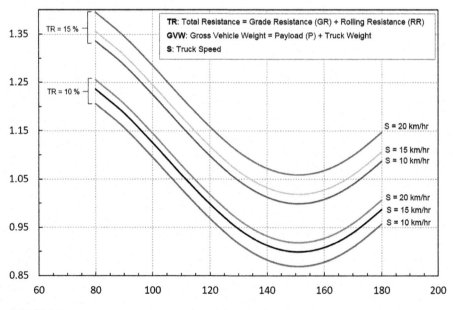

FIGURE 8.8
Correlation calculated by ANN between *P*, *S*, TR, and FC$_{Index}$ (CAT 777D).

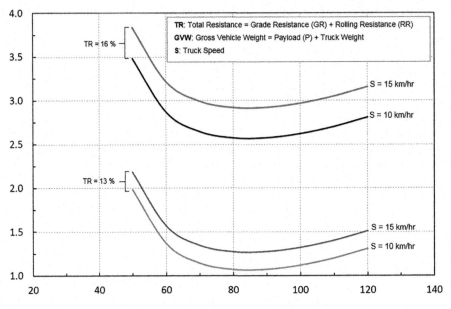

FIGURE 8.9
Correlation calculated by ANN between *P*, *S*, TR, and FC_{Index} (CAT 775G).

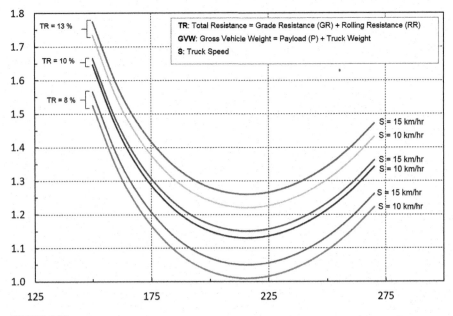

FIGURE 8.10
Correlation between *P*, *S*, TR, and FC_{Index} estimated by ANN (CAT 785D).

TABLE 8.4

Estimated Green House Gas Emissions by ANN (CAT 793D - Sample)

	(CO2-e) Index kg / (h.ton)					
	Total Resistance = 12%			Total Resistance = 8%		
GVW (ton)	S = 20 (km/h)	S = 15 (km/h)	S = 10 (km/h)	S = 20 (km/h)	S = 15 (km/h)	S = 10 (km/h)
340	0.599	0.589	0.581	0.559	0.556	0.554
360	0.591	0.586	0.578	0.556	0.554	0.549
380	0.589	0.583	0.575	0.554	0.551	0.548
400	0.589	0.583	0.575	0.554	0.551	0.548
420	0.591	0.586	0.578	0.556	0.554	0.551

and payload variability. Nevertheless, the connection between all essential factors and consumption of energy is not apparent. As a result, a different artificial intelligence model is needed to find the ideal value for the selected elements in order to reduce cargo FC.

The GHG emotions produced by haul trucks in surface mines can be calculated by the FC_{Index} predicted in the ANN model. Table 8.4 displays an average CAT 793D $(CO_2-e)_{Index}$ at the studied Australian mine site. The indicator shows how much CO_2-e is generated in an hour to push one ton of mining material.

The results show a logical link between the emissions of GHGs emitted and the truck activity parameters. The speed and total resistance of the truck would increase emissions of carbon. If the truck is transported by the supplier with the required payload, the minimum amount of gas is produced. With full cars, the fleet would significantly increase gas emissions.

Optimization of Efficient Mine Truck FC Parameters

Optimization

Optimization is a real way of identifying the best-quantified solution to problems as part of the quantitative work. Two elements must be considered in order to solve technical problems. The first is the field of research, and the second is a set of objectives. The research field considers every potential of the solution. The objective function is a math function that links each point in the answer field to a real value suitable for evaluating all field members. In mining engineering, solving numerous numerical problems was an on-going task. The stability of their mathematical models, which they challenge to portray in real complex and dynamic circumstances, characterize conventional optimization models. The implementation of artificial intelligence optimization techniques, such as heuristic search techniques, reduced the complexity of stiffness. Heuristic rules can be described well as applicable laws, which are the product of the experience and observation of behavioral system patterns in an experiment. They are suitable for solving all sorts of technical problems. In the 1950s, heuristic patterns were proposed using natural equivalences to model biological phenomena in engineering. These models are called natural methods of optimization. One of the significant benefits of using templates is a random function. In the 1980s, the use of these models to automate functions and processes was accomplished by improving the computers when traditional models were not successful in this area. Some new heuristic models, such as simulated annealing, swarm algorithms, ant colony optimization, and GAs, have been built in the nineties by previously completed algorithms.

Genetic Algorithms

As a biological engineering approximation, Holland (1975) proposed GAs to design and implement functional adaptive systems based on principles from both natural and genetic evolution [40]. GAs are relatively modern optimization models in the new generation. We use no quantitative information. We are thus very likely to escape a local minimum. Their use in similar technical issues leads to globally optimal solutions or, at least, more appropriate solutions than those obtained from other conventional mathematical models. They use a simple comparison of evolutionary processes. People from the field of research are picked by chance. The fitness of the answers resulting from the parameter to be optimized is subsequently determined by the fitness function. The person who produces the best fitness within the population has the highest opportunity to enter the next generation and repeat with another person by crossover and establish decadences with both characteristics. If an AG is appropriately developed, then the population (a group of possible solutions) converges to an optimal answer to the given problem. The methods most involved in evolution are crossover, based on the selection, reproduction, and mutation. In some technological, science, and economic problems, GAs have been used [38,40–43] due to their potential as optimization methods for multifaceted functions. There are four significant advantages when using GAs to solve problems. First of all, GAs do not have different optimization problems in mathematical requirements. Second, GAs can deal with certain kinds of objective functions and restrictions in discrete, continuous, or mixed fields of research. Third, the abundance of evolution operators makes GAs extremely useful for global searches. Finally, GAs give us great versatility to hybridize domain-dependent heuristics so that the problem can be efficiently applied. Apart from genetic operators, the influence of some variables on GA behavior and performance also needs to be analyzed in order to find them following the problem requirements and available properties. The effect of each factor on the efficiency of the algorithm depends on the class of issues being discussed. The determination of

TABLE 8.5

GA Parameters

GA Parameter	Details
Fitness function	The key optimization function
Individuals	A person is subject to any parameter for fitness.
Populations and generations	A population is a variety of people. In each iteration, the GA measures the existing population in order to produce a new generation.
Fitness value	The individual's fitness value is the value of each fitness function.
Children and parents	The GA selects specific individuals from the existing population, called parents, for the next generation.

an optimal value group of these variables will, therefore, depend on a large number of experiments and tests. In the GA system, there are many primary parameters. Table 8.5 displays the specifics of these critical parameters.

The key GA parameters are the population size, which affects the overall efficiency and performance of the GA. This mutation rate prevents a particular position from standing in value or from randomizing the quest.

GA System Developed

This chapter presents the GA model to boost three efficient critical energy consumption parameters of freight trucks at the mining site studied. The GA was chosen as an optimization technique mainly because it provides a variety of solutions and because of its parallel power during the search process. The main objective of this optimization method is to provide a set of P, S, and TR values for the final user, to achieve a minimum of FC_{Index}. This range of values is essential for real applications; for example, truck operators cannot hit an exact speed point or even an average for the whole cycle time.

A key point in using GA as an optimization mechanism is the management of population viability. Both individuals should be tested over generations if they are in the same range (i.e., maximum and minimum values) in which the ANN has been educated for two main reasons. First, the ANN only mapped the relationship between P, S, TR, and FC_{Index} based on data given during the training stage, and only in this distribution are predictions results and fitness values accurate. Second, each attribute's values must reflect the reality of mine sites and the limited operation of trucks to provide viable solutions.

In the built model payload, truck speed and total resistance are the individuals and mine truck FC is the primary function of optimizing. The fitness function was developed by the ANN algorithm in this model. Figure 8.11 shows all GA processes in the developed model.

Seven main processes were defined in this model. These procedures include initialized, encoded, crossed, mutated, decoded, selected, and replaced. Examples of the above procedures can be found in Table 8.6.

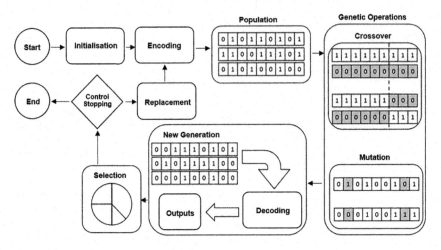

FIGURE 8.11
Processes of GAs (developed model) [14].

TABLE 8.6

GA Procedures

Procedure	Details
Initialization	Produce an original candidate solution population
Encoding	Digitalize population original value
Crossover	Combine two or more parental responses to make one new
Mutation	Chain of divergence. It is intended to rarely eliminate one or more people from a minimum local area and theoretically find a better answer
Decoding	Adjust a new generation's digital format to the original
Replacement	Replace people with excellent parental health

The key factors used to monitor algorithms in this established model were R^2 and MSE.

In this research, computer code in Python was written to complete the ANN and GA models created. Payload, truck speed, and total resistance are the algorithm input parameters in the first step (see Figure 8.12).

Based on the completed ANN model, the completed model generates a fitness feature. This role is the relation between FC and inputs. In the second step, the function formed takes the GA (optimization) process as an input to the computer code. The finalized codes start all GA procedures according to model-defined stopping criteria (MSE and R^2).

Finally, the algorithm presents structured parameters (*P*, *S*, and TR). These enhanced factors can be utilized to minimize the mine truck FC. In the built models, all procedures are based on existing data from a large surface mine in Australia. However, for other surface mines, the completed methods can be prepared by changing the data.

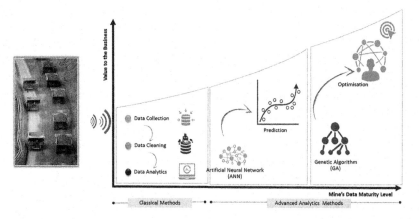

FIGURE 8.12
Final developed model.

Outcomes

The first step towards using the established optimization model is to define the range of all variables (individuals and minimum and maximum values). This range is calculated based on the collected data in the developed model. The parameters for controlling model generation are R^2 and MSE. Figure 8.13

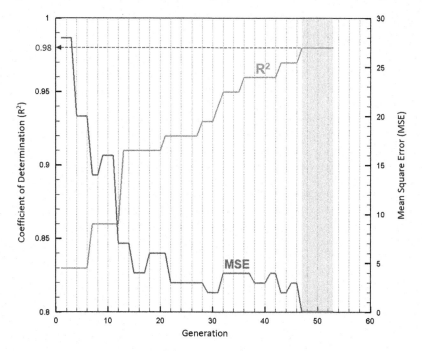

FIGURE 8.13
The determination coefficient and mean square error for all generations.

demonstrates the variability of these parameters in the studied mine site in generations.

In the studied mine site, R^2 was about 0.98, and after the forty-seventh generation, the MSE was about 0. These values were only updated until, in the fifty-third generation, the GA was stopped. The control factor values were thus stable after the forty-seventh generation, but the algorithm continued all procedures until the fifty-third generation. This is because the algorithm has established a confidence interval to obtain reliable results. Figure 8.14 shows the importance of the fitness function (FC_{Index}) in all generations. The estimated FC of mine truck ranges from 0,03 to 0,13 (L/ (h. ton)). The calculated results have a mean of 0.076 (L/ (h. ton)), and over 45% of the results are above the average. Although the model presented could lead to local FC minimization, acceptable results can be found after the forty-seventh generation. Figure 8.14 also shows that the FCIndex is approximately 0.04 (L/ (h. ton)) in the appropriate area. This means that the minimum FC_{Index} for the CAT 793D is approximately 0.04 (L/ (h. ton)) by increasing weight, truck speed, and total strength at the studied mine site.

In these case studies (all mine sites), the optimal set of variables to reduce FC is described in Table 8.7.

Besides, the mine managers approved the use of the proposed optimization model for the studied mine site, which decreased FC by 9% and related GHG emissions by six months with the production application in the mine. The management team also recorded a 5% increase in efficiency during the application testing period.

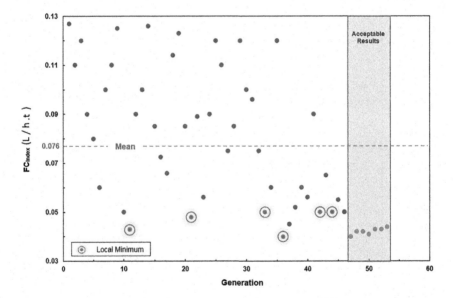

FIGURE 8.14
FC in all generations.

TABLE 8.7

Recommendations for Optimization Models to Optimize Energy Efficiency Gains

Truck	Variables	Normal		Optimized	
		Min	Max	Min	Max
CAT 793D	Gross vehicle weight (ton)	150	380	330	370
	Total resistance (%)	8	20	8	9
	Truck speed (km/h)	5	25	10	15
CAT 777D	Gross vehicle weight (ton)	65	150	145	150
	Total resistance (%)	9	25	9	11
	Truck speed (km/h)	10	45	10	12
CAT 775G	Gross vehicle weight (ton)	45	95	75	90
	Total resistance (%)	13	20	13	14
	Truck speed (km/h)	5	55	9	13
CAT 785D	Gross vehicle weight (ton)	125	215	200	215
	Total resistance (%)	8	15	8	9
	Truck speed (km/h)	5	45	10	15

Conclusion

This chapter and algorithms aimed at improving energy efficiency and reducing emissions of GHG by looking at specific databases based on the correlations between effective parameters. These relationships were complex, and the techniques of artificial intelligence were necessary to establish a simple algorithm to tackle this problem. A relationship between the examined parameters has been explained in the first section of this chapter by an ANN algorithm. The results showed a nonlinear association with the parameters tested in FC. The ANN has been taught and checked by the collected datasets from actual mine sites. The achievements showed that the calculated and real values of the truck intake of fuel were well organized. In the second part of this chapter, a GA was developed to reduce FC in transport operations. The results show that the optimization of effective energy consumption factors can be achieved by applying this approach. The algorithm developed could detect local fitness minimum levels. The GA model provided underlined positive results in order to reduce the rate of surface mining FC. The range of efficient FC parameters were optimized for haulers and the best payload, truck speed, and road resistance values were underlined to minimize the FC_{Index}. The models can be enhanced by increasing the number of input parameters in this chapter. In real mine sites, selected parameters can be tested. Nonetheless, it is a bit difficult for current systems to adjust the total resistance and monitor the variation in the payload. The algorithms mentioned in this chapter can replace other achievable parameters, including idle time, queuing time, etc. with model parameters.

References

1. Golosinski, T., Mining education in Australia: A vision for the future. *CIM Bulletin*, 2012. 93(1039): pp. 60–63.
2. Zheng, S. and H. Bloch, Australia's mining productivity decline: Implications for MFP measurement. *Journal of Productivity Analysis*, 2014. 41(2): pp. 201–212.
3. DOE, Mining industry energy bandwidth study. 2012: Department of Energy, USA Government, Washington DC. pp. 26–33.
4. Kecojevic, V. and D. Komljenovic, Haul truck fuel consumption and CO^2 emission under various engine load conditions. In *SME Annual Meeting and Exhibit, CMA 113th National Western Mining Conference*, Colorado Convention Center, Denver, CO, USA 28 Feb – 2 Mar 2011: SME. pp. 186–195.
5. Soofastaei, A. and J. Davis, Advanced data analytic: A new competitive advantage to increase energy efficiency in surface mines. *Australian Resources and Investment*, 2016. 1(1): pp. 68–69.
6. Beatty, J. and D. Arthur, Mining truck operations. In *Mining Truck Operations in Australia*. 1989: AusIMM Bulletin, Melbourne.
7. Beckman, R., Haul trucks in Australian surface mines. *Australia*, 2012. 1: pp. 87–96.
8. DOE, Energy and environmental profile of the US mining industry. 2002: Department of Energy, USA Government, Washington DC. pp. 63–87.
9. Carmichael, D., B. Bartlett, and A. Kaboli, Surface mining operations: Coincident unit cost and emissions. *International Journal of Mining, Reclamation and Environment*, 2014. 28(1): pp. 47–65.
10. Cetin, N., Open-pit truck/shovel haulage system simulation. A Thesis of the Graduate School of Natural and Applied Sciences of Middle East Technical Universality. Turkey, 2004. 1(2): pp. 147–156.
11. Coyle, M., Effects of payload on the fuel consumption of trucks. 2007: Department for Transport, London. pp. 36–40.
12. Soofastaei, A., et al., Payload variance plays a critical role in the fuel consumption of mining haul trucks. *Australian Resources and Investment*, 2014. 8(4): pp. 63–64.
13. Soofastaei, A., et al., Energy-efficient loading and hauling operations. In Awuah-Offei, K. (ed.) *Energy Efficiency in the Minerals Industry*. 2018: Springer, Cham. pp. 121–146.
14. Soofastaei, A., et al., A discrete-event model to simulate the effect of truck bunching due to payload variance on cycle time, hauled mine materials and fuel consumption. *International Journal of Mining Science and Technology*, 2016. 26(5): pp. 745–752.
15. Soofastaei, A., et al., A comprehensive investigation of loading variance influence on fuel consumption and gas emissions in mine haulage operation. *International Journal of Mining Science and Technology*, 2016. 26(6): pp. 995–1001.
16. Soofastaei, A., et al., Development of a multi-layer perceptron artificial neural network model to determine haul trucks energy consumption. *International Journal of Mining Science and Technology*, 2016. 26(2): pp. 285–293.
17. Soofastaei, A., et al., Reducing fuel consumption of haul trucks in surface mines using artificial intelligence models. In *Coal Operators' Conference*. 2016: University of Wollongong, New South Wales, pp. 477–489.

18. Soofastaei, A., et al., Development of an artificial intelligence model to determine trucks energy consumption. In *Energy Future Conference*. 2014: Future Energy, University of New South Wales, Sydney, pp. 178–179.

19. Soofastaei, A., et al., The influence of rolling resistance on haul truck fuel consumption in surface mines. *Tribology International Journal*, 2016. 2(1): pp. 215–228.

20. De Francia, M., et al., Filling up the tank. *Australasian Mining Review*, 2015. 2(12): pp. 56–57.

21. Alarie, S. and M. Gamache, Overview of solution strategies used in truck dispatching systems for open pit mines. *International Journal of Surface Mining, Reclamation and Environment*, 2002. 16(1): pp. 59–76.

22. Bhat, V., A model for the optimal allocation of trucks for solid waste management. *Waste Management and Research*, 1996. 14(1): pp. 87–96.

23. Burt, C.N. and L. Caccetta, Match factor for heterogeneous truck and loader fleets. *International Journal of Mining, Reclamation and Environment*, 2007. 21(4): pp. 262–270.

24. Nel, S., M.S. Kizil, and P. Knights, Improving truck-shovel matching. In *35th APCOM Symposium*. 2011: Australasian Institute of Mining and Metallurgy (AusImm), University of Wollongong, New South Wales, pp. 381–391.

25. Caterpillar, *Caterpillar Performance Handbook*, 10 edition. Vol. 2. 2013: US Caterpillar Company, New York.

26. Caterpillar, CAT 793D mining truck, CAT, Editor. 2013: Caterpillar, New York. pp. 4–7.

27. Filas, L., *Excavation, Loading and Material Transport*. Vol. 2. 2002: Littleton Co.

28. Runge, I.C., *Mining Economics and Strategy*. Vol. 4. 1998: Society for Mining, Metallurgy, and Exploration, Sydney.

29. ANGA, National greenhouse accounts factors, I. Department of Industry, Climate Change, Science, Research and Tertiary Education, Editor. 2013: Australian Government, Canberra, pp. 326–341.

30. Kecojevic, V. and D. Komljenovic, Haul truck fuel consumption and CO_2 emission under various engine load conditions. *Mining Engineering*, 2010. 62(12): pp. 44–48.

31. Kecojevic, V. and D. Komljenovic, Impact of bulldozer's engine load factor on fuel consumption, CO_2 emission and cost. *American Journal of Environmental Sciences*, 2011. 7(2): pp. 125–131.

32. DCE, Emission estimation technique manual. 2012: The Department of Climate Change and Energy Efficiency, Australian Government, Canbera, Australia. pp. 126–141.

33. Hammood, A., Development artificial neural network model to study the influence of oxidation process and zinc-electroplating on fatigue life of gray cast iron. *International Journal of Mechanical and Mechatronics Engineering*, 2012. 12(5): pp. 128–136.

34. Xiang, L., Y. Xiang, and P. Wu, Prediction of the fatigue life of natural rubber composites by artificial neural network approaches. *Materials and Design*, 2014. 57(2): pp. 180–185.

35. Sha, W. and K. Edwards, The use of artificial neural networks in materials science based research. *Materials and Design*, 2007. 28(6): pp. 1747–1752.

36. Talib, A., Y. Abu Hasan, and N. Abdul Rahman, Predicting biochemical oxygen demand as indicator of river pollution using artificial neural networks. In *18th World Imacs/Modsim Congress*. 2009: Cairns, Australia. pp. 195–202.

37. Ekici, B. and T. Aksoy, Prediction of building energy consumption by using artificial neural networks. *Advances in Engineering Software*, 2009. 40(5): pp. 356–362.
38. Beigmoradi, S., H. Hajabdollahi, and A. Ramezani, Multi-objective aero acoustic optimisation of rear end in a simplified car model by using hybrid robust parameter design, artificial neural networks and genetic algorithm methods. *Computers and Fluids*, 2014. 90: pp. 123–132.
39. Rodriguez, J., et al., The use of artificial neural network (ANN) for modeling the useful life of the failure assessment in blades of steam turbines. *Engineering Failure Analysis*, 2013. 35: pp. 562–575.
40. Lim, A.H., C.-S. Lee, and M. Raman, Hybrid genetic algorithm and association rules for mining workflow best practices. *Expert Systems with Applications*, 2012. 39(12): pp. 10544–10551.
41. Reihanian, M., et al., Application of neural network and genetic algorithm to powder metallurgy of pure iron. *Materials and Design*, 2011. 32(6): pp. 3183–3188.
42. Velez, L., Oswaldo, genetic algorithms in oil industry: An overview. *Journal of Petroleum Science and Engineering*, 2005. 47(1): pp. 15–22.
43. Opher, T. and A. Ostfeld, A coupled model tree (MT) genetic algorithm (GA) scheme for biofouling assessment in pipelines. *Water Research*, 2011. 45(18): pp. 6277–6288.

9

Making Decisions Based on Analytics

Paulo Martins and Ali Soofastaei

Introduction

The Big Data (BD) era and the power of Advanced Analytics (AAs) tools are fundamentally changing the way companies make decisions, disrupting existing business models and ecosystems. Advances in digital and mobile technology, new computer architecture, and intelligent software have resulted in a vast proliferation of data and allowed us to collect, store, and process data from almost anything in different forms and magnitude. The power created by this never-before-seen technological scenario raises some critical questions:

- what to do with all this information and how to use it for minimizing risks and making better decisions?
- considering all sophisticated analytical tools, how to incorporate their insights into the organizational culture?

Traditional successful leaders most commonly rely on their experience and intuition to make crucial decisions. New generations of digital leaders face real challenges to put their transformational view and guide companies towards data-driven objectives. They must deal with several variables, and the speed and volume of transactions are just too much for human decision-makers.

In this regard, Stanford professors Jeffrey Pfeffer and Robert Sutton explored the gap between having information and acting. The authors explained how some action inhibitors such as exaggerated competition between internal departments, inefficient and complex performance measurement systems, and lack of critical thinking could undermine an organization's ability to progress despite having all this knowledge available to them. Unfortunately, turning awareness into doing, with actionable initiatives, is hard, creating an ever-widening gap between leaders who are acting and those who are not [1].

In this landscape, AAs surge as a potential agent to help companies redesign their strategies and enhance their capacity to put data into action. The companies have obtained real value in conjunction with real-time data updates from predictive and prescriptive analytic techniques to enable dynamic decision-making. Industry 4.0 emerging knowledge such as natural language processing, machine learning (ML), automation, and intelligent machines can create leaps of efficiency, meaning, and insight that is hidden within the business and the world at large – a unique opportunity for leaders to make more informed and effective decisions [2].

Indeed, this is simultaneously the best and the worst of times for decision-makers. This is best because of the considerable amount of affordable data coupled with improved analytics methods and a better knowledge of how to mitigate the cognitive biases that often undermine the decision process in an organization. This is the worst because of the frequent disconnection between organizational dynamics and digital decision-making increases the levels of frustration among operational managers and executive leaders [3].

While the concept of BD and AAs has been around for years, many companies are still lagging when it comes to incorporating data to improve the decision-making process. According to the Global Data and Analytics Survey 2016: Big Decisions from PWC, more than two-thirds (61%) of participants say their business decisions are only somewhat or rarely data-driven [2]. A recent Deloitte survey reported that although many companies have adopted technologies and created data teams, few are making the transition to reach higher maturity levels and fully embrace data analytics. The report correlated the adoption of a fragmented, siloed strategy to data analytics techniques and technologies, with a decline in the success rate achieved by many enterprises, and highlighted the importance of culture-orientation to data-driven responsibility assignment, adequate training, and leadership in becoming a data-driven organization. Among those, culture was identified as the most common culprit, and only 39% of respondents consider their company has embedded a powerfully built cultural orientation to benefit from data insights. Additionally, the survey showed that only 37% of respondents believe that the workforce in their company understands the importance of data.

Buying and using analytics tools is not hard— changing behaviors is [4].

Surprisingly, the survey showed that 67% of those surveyed express discomfort consulting or manipulating data from their available resources and tools, an alarming statistic considering that data and analytics are at the heart of smart business decisions. Current leaders need to identify data quality issues and develop an enterprise-wide approach to ensure that information is being extracted or created on an ongoing basis. This involves drawing a detailed road map for investing in technology and building a common and standardized baseline for monitoring and managing risk [4].

The digital revolution is disrupting the way managers use data to improve several operations, including logistics, resource allocation, scheduling, manufacturing, maintenance planning and execution, marketing, and sales. Many organizations are following these trends and integrating continuous improvement, innovation, business intelligence, and operations research in the same environment of AAs.

Comprehensive analysis can significantly improve decision-making, reduce risk, and uncover valuable insights from data that would otherwise remain hidden [5]. As BD has increasingly scaled up, AAs tools become more and more essential to enable understanding, find patterns, and optimize processes. Statistical and mathematical techniques are being primarily adopted to address business problems and systemize decision-making. With advanced statistical models, valuable insights are obtained, and companies can explore new opportunities. For example, managers can use data analytics to boost their efficiency when faced with complicated financial dashboards and reports. Analytical methods can be applied to identify trends and extract detailed insights to evaluate cost variations, understand competitors' position, and define customer and product segmentation strategies. Similarly, AAs calculation can be applied to solve business challenges in operational excellence, product development, and strategic planning areas.

Analytics has also proved to be a high-value approach for strategic decisions in the supply chain area. Inefficiencies can be identified, and improvement opportunities unhidden. Risk modeling and assessments can be executed to support investment decisions. Some other improvement opportunities include inventory management, channel management, procurement, and logistics [6].

Organization Design and Key Performance Indicators (KPIs)

Organizational Changes in the Digital World

The accelerated pace of strategic change driven by disruptive technologies has resulted in an increasing trend for companies to review their organizational structures. To boost a company's performance, business leaders have focused on their ability to adapt to a more complex environment defined by hyper-competition, disruptive technologies, high customer standards, and tight regulation. The introduction of technologies such as automation and ML/artificial intelligence (AI) can dramatically change how work gets done – they alter some processes and eliminate others. Companies that prioritize their digital journey are committed to rewiring their organizational structure to reflect their objectives and capture all the potential benefits.

Despite the growing trend, the statistics related to reorganization initiatives are not favorable. Surveys conducted by top management firms reported that more than 80% of the respondents indicated that they had experienced a recent redesign in their organizations, and less than 50% rated fewer than half of the reorganization efforts as successful [7,8]. To successfully reach the desired result, leaders must deeply work on different aspects such as hierarchy, number of layers, leadership and talent capabilities, performance-management system, and communication. In today's fast-changing world, many companies have replaced their traditional organizational structure and successfully embraced Agile principles to establish a modern and adaptable operating model.

Traditional organizational structures, based on most of Taylor's and Ford's principles associated with Scientific Management, have dominated industries from 1911 to 2011, known as "the management century." This model opened an era of unprecedented effectiveness and efficiencies leveraged by some highly adopted principles, such as Quality Control, Total-Quality Management, Knowledge Management, Lean Production, and Management [9,10].

The digital revolution is transmuting economies, industries, and societies. This quickly evolving environment, highly influenced by the constant introduction of technology and accelerated democratization of information, demands more flexibility and capacity to react to changes.

Traditionally, organizations are designed for and measured by stability in their structures, responsibility profiles, and goal systems. Their fixed and vertical hierarchy structure – box and lines on the org chart – typically defines where activities are executed, how performance is measured, and who is responsible for awarding bonuses. Usually, a boss oversees the team and manages direct reports. Business leaders believe their organizations should operate in a mathematical model where structure, governance, and processes are aligned to perform in a clear, predictable way [11]. The decision-making process in this type of structure follows a top-down hierarchy for almost all levels of decision, slowing down the flow.

On the other hand, Agile organizations first define the "primary" one dimension of their organizational structure and, consequently, where individual workers work. Performance monitoring, and determination of benefits, however, are most likely to happen in teams that cut across formal structures. Supported by a backbone of stable elements, an iterative and flexible architecture is followed. They master the art of stability and flexibility and can quickly respond to new market opportunities and deploy resources where need. With a focus on rapid interaction and experimentation, they are open to keep continually learning and face uncertainty [12]. To become more agile, companies are modifying their organizational structure, redefining roles, and improving integration between the business and information technology. Agile management practices promote continuous improvement, adaptive planning, evolutionary development, innovation, collaboration, and rapid and flexible response to environmental conditions, market, customers, technology, and competition [13].

Embedding KPIs in the Organizational Culture

Regardless of whether it is based on Traditional or Agile concepts, organizational redesign comes together with an in-depth review of the performance management system, which includes governance mechanisms, business metrics, and measurement systems and methods. Top-down strategies that cascade down companies' targets – created to achieve chief executive officer's and senior leaders' targets – to business units, smaller units, and ultimately individuals, may, along the way, produce a siloed and dysfunctional group of metrics. To overcome this problem, common KPIs must be clearly defined [11]. By incorporating KPIs, companies can change organizational practices and interdependency between methods. They are a powerful tool to align different divisions and individuals with one goal, making performance management all the members' responsibility.

Although some studies have shown that performance measurement practices can shape the employee's mindset and behavior and vice versa, the cultural aspect is not commonly explored in the literature [14]. KPIs are the key to development, and an effective performance management system and successful implementation require a stable integration among departments and business units combined with increased perception of its importance. Business leaders are responsible for ensuring the correct definition of KPIs to support business strategies.

In the current practice, dashboards are used to report and present KPIs with contextual information to help the identification of deviations and their root causes. However, some drawbacks are related to the complexity of finding relationships between KPIs and their characteristics [15]. Many companies face the challenge of standardizing KPI definition and data granularity level across all parts of the business. A classic example is the cost allocation definition, where some areas allocate items as variable costs while others define them as fixed costs. A consensus around this common taxonomy is essential to succeed.

Performance-management systems have significantly evolved over the years, and the metrics they use have become more complex and sophisticated. KPIs are not always well defined and communicated, and sometimes it is hard to define an appropriate KPI to connect with each business objective. New sources of data, in different forms (e.g., unstructured data) and increased sizes, have the potential to enhance the company's analytical capabilities by reinforcing their association with internal KPIs and creating value by making these data actionable [16]. Application of data analytics takes the performance-management cycle to a higher level since it provides advanced insights and pattern identification to complement the simple visualization provided by traditional Balanced Score Cards and descriptive dashboards.

This is much more than metrics management. Indeed, data analytics creates a healthy skeletal structure that enables continuous improvement and optimization. It helps to investigate what is working and what is not working in each department. High-performance organizations can even automate several decision processes, for example, using Robotic Process Automation to redesign routine tasks and establish intelligent systems to control process variability. Relevant data extracted from internal and external sources can be used to build predictive models aimed to forecast gaps on strategic KPIs, guiding decision-makers towards the company's objectives.

For example, the AAs firm, Quantum Black, developed a predictive model using the operational data from a mining company such as operating hours, ore characteristics, and load volume and external data like weather conditions. The analytical model can detect the initial signs of failures, such as motor voltage, current, and temperature, empowering the company's capacity to identify and solve persistent failures on its conveyor belts [17].

The next sections will present a set of tools and methods based on AAs technologies designed to leverage the decision-making capabilities of people from different levels in an organization.

Decision Support Tools

Despite the importance of AAs, many companies are still struggling to incorporate those principles in the corporate culture and make the best use of its insights for decision-making. Leaders are continually seeking to understand the mechanisms to make more informed decisions and desire efficient ways to access sophisticated ideas so they can quickly absorb the meaning behind the data and efficiently act.

New tools based on AAs can enhance the traditional decision-making process, transforming data into information and insights, uncovering patterns previously invisible and presenting them in a ready-to-use format. What-if scenarios, simulations, and predictions executed sometimes in real time can provide immediate guidance in an increasingly dynamic environment [18]. Optimization algorithms based on AI methods are being developed to deal with highly complex problems, such as energy efficiency [19] and logistics [20,21].

Making informed decisions in the current disruptive scenario includes knowing how to handle BD effectively and how to apply the available resources required to explore the maximum benefit (Figure 9.1).

A four-layer hierarchy known as the Data, Information, Knowledge, and Wisdom model is useful to describe methods for problem-solving and decision-making. The most basic level is Data, followed by Information, which

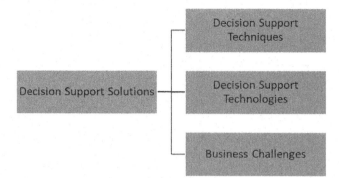

FIGURE 9.1
Components of decision support solutions.

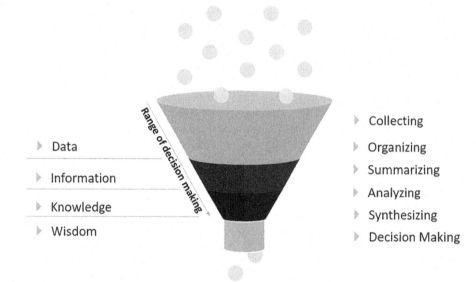

FIGURE 9.2
Data, Information, Knowledge, and Wisdom hierarchy.

adds value extracted from data; afterward, Knowledge adds how to use data, and finally, Wisdom adds when and why to use the acquired data (Figure 9.2).

In the context of BD Analytics, decision-making tools can be divided into two groups: decision-making techniques and decision-making technologies. Regardless of the business challenge, many techniques and technologies have been developed and can be combined to effectively support each phase of processing, analyzing, visualizing, and communicating BD results [22] (Figure 9.3).

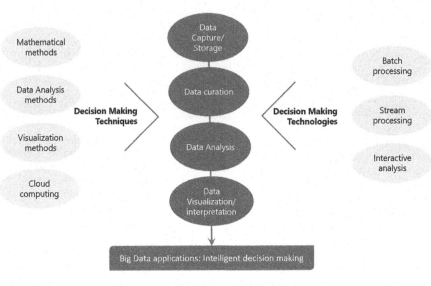

FIGURE 9.3
Techniques and technologies to support the analytics of BD [22].

Decision-making techniques – They are developed to make decision-making possible in each phase of the analytical process. Several methods are applied based on many disciplines; some categories are mathematical tools, data analysis tools, visualization tools, and other more sophisticated techniques. Some examples of those techniques are described in Figure 9.4. The models do not fully extend to all their subfields but provide an overview of some relevant methods.

Decision-making technologies – They involve a set of new technologies designed to deal with the challenges of BD (storage and processing). Batch processing was the first generation of technologies and only considered data that are already in the data storage. Scalability and reliability are the main advantages of batch processing; however, a limitation is found in terms of high-throughput latency in its implementation. To tackle this challenge, stream processing was developed. It is designed for real-time analytics typically supported by a distributed environment. The third generation is a hybrid processing system that synthesizes batch processing and streaming based on the Lambda Architecture in three different layers (batch layer, speed layer, and serving layer).

A description of some processing technologies developed for each paradigm is presented in Figure 9.5.

To provide a robust framework to support the integration of BD analytics in the business decision-making process, [5] presented a five phases

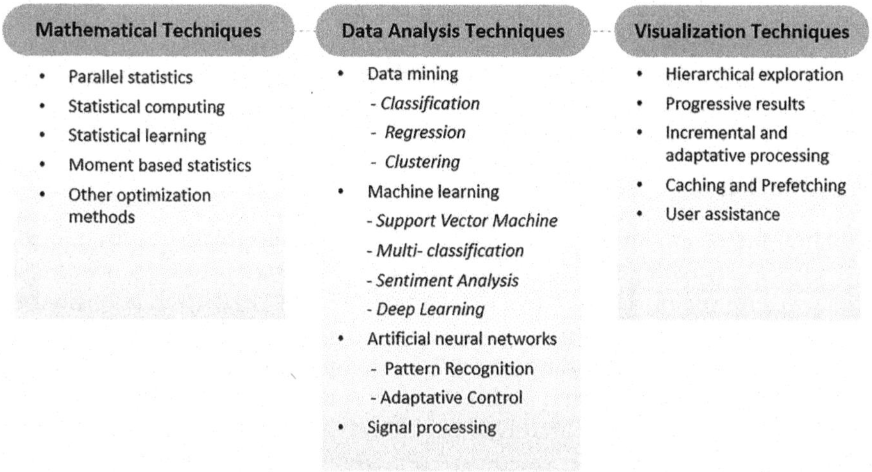

Mathematical Techniques	Data Analysis Techniques	Visualization Techniques

• Parallel statistics
• Statistical computing
• Statistical learning
• Moment based statistics
• Other optimization methods
	• Data mining
	  - *Classification*
	  - *Regression*
	  - *Clustering*
	• Machine learning
	  - *Support Vector Machine*
	  - *Multi- classification*
	  - *Sentiment Analysis*
	  - *Deep Learning*
	• Artificial neural networks
	  - Pattern Recognition
	  - Adaptative Control
	• Signal processing
		• Hierarchical exploration
		• Progressive results
		• Incremental and adaptative processing
		• Caching and Prefetching
		• User assistance

FIGURE 9.4
Examples of decision-making techniques [22,23].

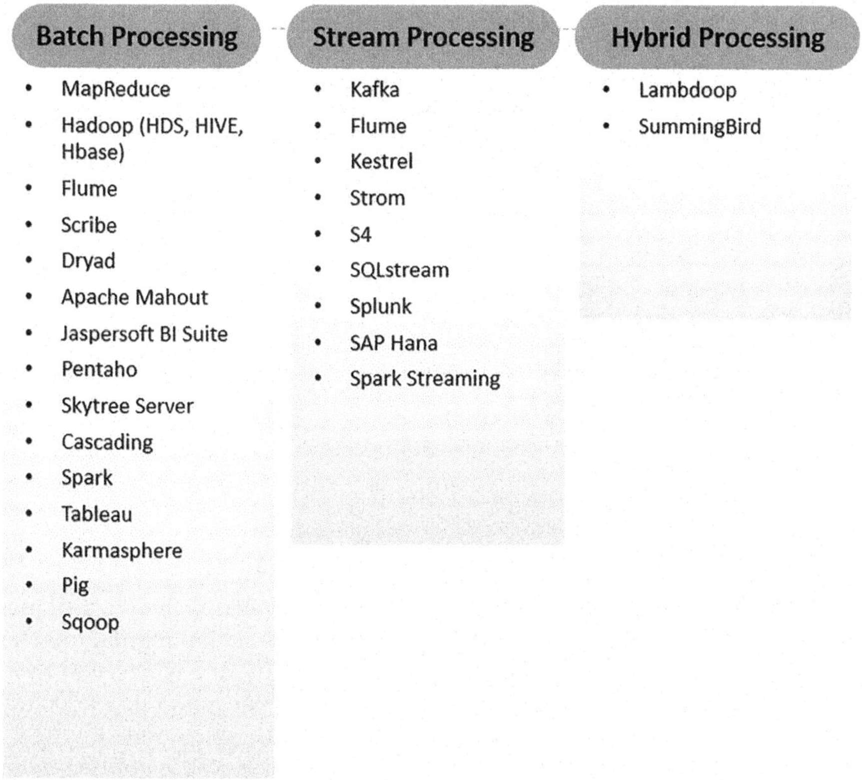

Batch Processing	Stream Processing	Hybrid Processing
• MapReduce • Hadoop (HDS, HIVE, Hbase) • Flume • Scribe • Dryad • Apache Mahout • Jaspersoft BI Suite • Pentaho • Skytree Server • Cascading • Spark • Tableau • Karmasphere • Pig • Sqoop	• Kafka • Flume • Kestrel • Strom • S4 • SQLstream • Splunk • SAP Hana • Spark Streaming	• Lambdoop • SummingBird

FIGURE 9.5
BD technologies classified by distinct processing paradigm.

approach. The research explored some BD tools (technologies), techniques, and architectures as a conceptualization of some possible approaches to performing BD analytics. A summarized description of each phase is provided in the following.

Phase 1 – Intelligence

The first phase englobes the identification of different data sources and requisites for extraction, processing, storage, and migration to the end-user. In this phase, strategic data is collected, which includes structured and unstructured data gathered from machine outputs such as log files, sensor data, and mobile, GPS, and satellite data. In addition to operational data, there is social media data, text, images, and audio. After acquisition, this data can be appropriately stored in a traditional Database Management System – like MySQL or PostgreSQL – or NoSQL database – like MongoDB, CouchDB.

Phase 2 – Data Preparation

In this phase, the data is organized, prepared, and processed. High-speed networks using ETL/ELT or BD tools like Hadoop and MapReduce can be applied. For data query, computation, and processing a range of different programming languages are available, such as R, Python, Scala, and SQL. Some tools and platforms can facilitate the storage, manipulation, and data discovery activities in an integrated and single solution. Examples include SAP HANA, Vertica, and IBM Netezza.

Phase 3 – Design

The objective of this phase is to analyze possible courses of action considering a representative or conceptualization model of the problem. In this phase, three measures are described:

- Model Planning,
- Information Analytics, and
- Analysis.

In the planning stage, models and algorithms are selected. AAs techniques, such as clustering, classification, regression, and association regulations, can be applied and combined with AI methods and ML such as decision trees, neural networks, and pattern-based analytics. Further possibilities include time series, text analysis, graph analyses, and others.

Afterward, in the data analytics step, the selected model is applied using analytical tools and technologies such as RapidMiner, KNIME, HANA, Revolution R Enterprise, or even a created from the scratch algorithm in

Python. Finally, in the analyzing step, the outputs of the previous steps are analyzed, and the possible courses of action to be taken are defined to feed the next phase.

Phase 4 – Choice

In this phase, methods for evaluation of the impacts of each solution are applied. It is separated into two stages: assess and make a decision. In the evaluation stage, the proposed courses of actions and their impact are evaluated and prioritized using some tools such as reports, dashboards, simulation, what-if scenarios, KPIs, cognitive maps, and advanced or interactive data visualization solutions (e.g., Power BI, Tableau, Qlik View, and Spotfire). Subsequently, based on the results of the evaluation step, the decision is made choosing the best or most appropriate solution.

Phase 5 – Implementation

In the final phase of the decision-making process, the selected solution will be implemented, and its results are operationalized or put to action. Continuous monitoring can be established to provide real-time and regular feedback about the results of the decision.

The five phases presented above are a valuable resource to help decision-makers understand the current and emerging technologies and structures necessary to take advantage of the vast amount of data produced in their operations and available in external sources. In the next topic, the focus is on a more detailed description of some analytics applications designed to process individual data sets to support decision-making in specific fields.

AAs Solutions Applied for Decision-Making

Decision-making solutions developed around the world are created based on the efficient combination of techniques, technologies, and business rules and necessities. Most of the current solutions are disrupting traditional methods and boosting companies' performance. In the following, some of them are presented.

Intelligent Action Boards (Performance Assistants)

After all the hard effort to define relevant metrics and create an organizational structure to monitor and report performance of KPIs, it is expected that managers know what to do. However, dashboards are deluging administrators with much data and presenting too many actionable understandings, resultant in information paralysis [24]. The disconnection between theory and practice is not found in the concept of dashboard creation. Well deployed dashboards reflect companies' performance, are periodically updated – sometimes in near real-time – with accurate data, and are available for decision-makers. The problem is that with so much data to analyze, the managers' available time or skills to interpret the metrics is not enough. Consequently, there is a delay in the response time to prevent performance problems or actions are not taken. Additionally, opportunities to recognize strong performance may be missed.

Mckinsey and Systems Solutions Center created "Intelligent Action Boards" to help managers get things done [24]. Instead of only report metrics, "this next-generation instrument allows businesses to improve performance across a wide variety of roles, from companies to sales." The solution is designed to apply AAs to automatically extract insights from metrics using rules-based logic and configure alerts that tell decision-makers a clear course of actions (recommendation system) that are necessary to improve performance. Action boards will significantly increase the capacity of companies to take action and respond to their information. Figure 9.6 shows an example of an action board report for mining fleet performance. In the example, a predictive

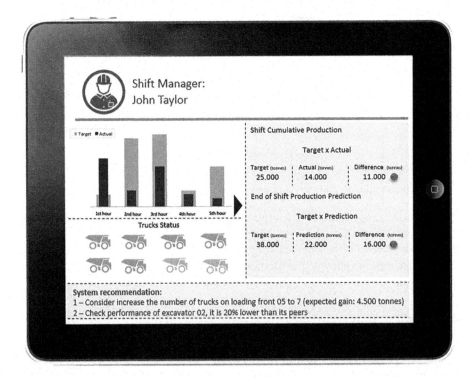

FIGURE 9.6
Smart action board implemented to manage mining fleet performance.

model forecasts the production by the end of the shift and performs AAs insight to deliver customized recommendations to the shift manager in order to improve production. Two recommendations are shown at the bottom of the report in this scenario.

Predictive and Prescriptive Models

Predictive models are developed to predict to help a person or organization make data-driven decisions by using data mining and the possibility to predict expected results. According to [25], "Predictive data analytics is the art of developing and applying methods which create forecasting based on models obtained from historical data." Forecast values using predictive models involve several independent parameters (predictors) that possibly affect the dependent parameter (target). Relevant variables are previously defined, and their data is submitted to a statistical algorithm, which may be a simple linear regression or an advanced ML algorithm, such as the Artificial Neural Networks (ANNs).

All predictive data analytics projects start with the business problem and the kind of insight that a predictive model can provide to help the organization solve the problem. It is crucial to be aware that predictive models are intended to empower the decision-making process. Therefore, they have a supportive function and are not meant to find the optimal decision. For this objective, prescriptive or optimization models are the best alternatives.

Prescriptive analytics is considered a key driver for companies to reach a higher level in the BD analytics journey. All levels of strategic decisions and opportunities to enhance business performance can be improved [26]. Its objective is utilizing the vast and diverse amount of available data to make predictions and recommend (prescribe) the best alternative. For developing this idea, predictive analytics output is incorporated and combined with AI, optimization methods, and intelligent platforms and methods in a probabilistic context to deliver flexible, constrained, automated, time-dependent, and optimized outcomes [26].

The way prescriptive analytics participate in the decision-making process may be classified into two levels: decision support – providing a direct recommendation or advanced insights, or decision automation – automatically implementing the prescribed action. The quality and applicability of the prescriptions are a result of the correct combination of structured and unstructured data, the coherence to represent the knowledge about the field under study, and the capacity to extract benefits of the decisions in question [26].

There are many different types of predictive and prescriptive modeling methods. The classification used here is based on the definition provided by [26], who identified three categories for predictive analytics: ML, Statistical Analysis, and Probabilistic Models; and six categories for prescriptive analytics: Probabilistic Models, ML/Data Mining, Mathematical Programming, Evolutionary Computation, Simulation, and Logic-based Models (Figures 9.7 and 9.8).

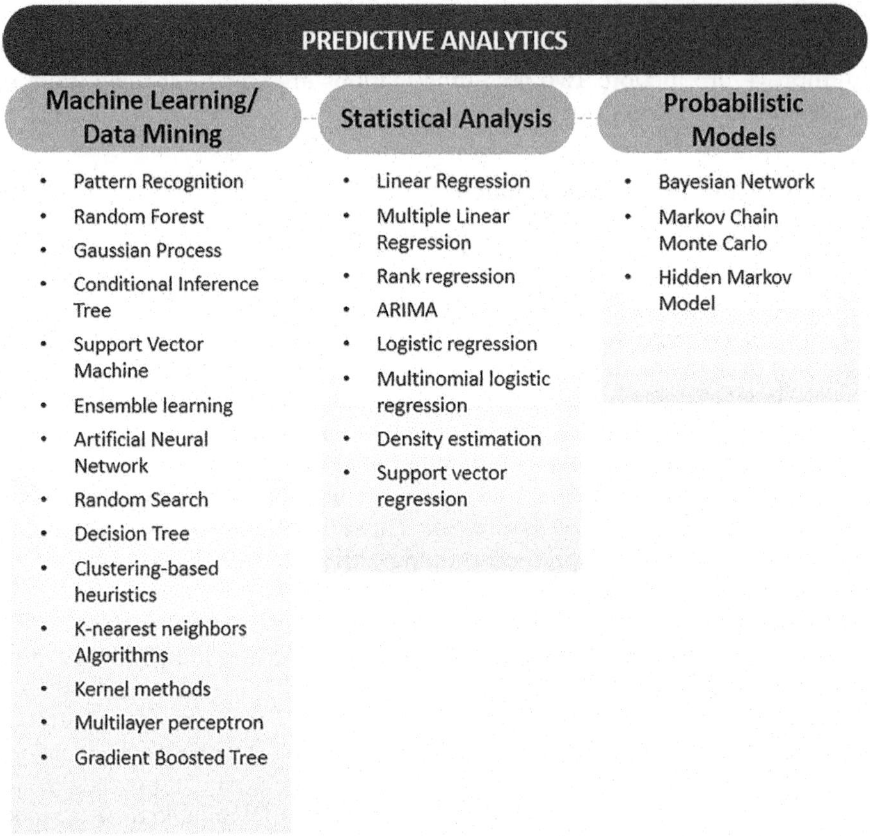

PREDICTIVE ANALYTICS

Machine Learning/ Data Mining	Statistical Analysis	Probabilistic Models
• Pattern Recognition • Random Forest • Gaussian Process • Conditional Inference Tree • Support Vector Machine • Ensemble learning • Artificial Neural Network • Random Search • Decision Tree • Clustering-based heuristics • K-nearest neighbors Algorithms • Kernel methods • Multilayer perceptron • Gradient Boosted Tree	• Linear Regression • Multiple Linear Regression • Rank regression • ARIMA • Logistic regression • Multinomial logistic regression • Density estimation • Support vector regression	• Bayesian Network • Markov Chain Monte Carlo • Hidden Markov Model

FIGURE 9.7
Classification of predictive analysis techniques [26].

Optimization Tools

There is a deep relationship between optimization and prediction models. Hertog and Postek mentioned that "optimization is utilized to achieve predictive patterns, and predictive devices are used to predict variables in optimization methods" [27]. Optimization methods are widely used in many industries. Most of the applications are intended to optimize productivity, energy and cost efficiency, and safety. Before the advent of BD, the optimization methods traditionally used were designed to develop practical solutions for business problems. The latest developments allow us to access, store, and process high-quality data in large magnitudes, creating an excellent opportunity to use groundbreaking optimization methods to achieve better results. Among all existing methods, particle swarm, genetic algorithm, bee colony, ant colony, firefly algorithm, and tabu search are the most significant relevant in critical businesses [28].

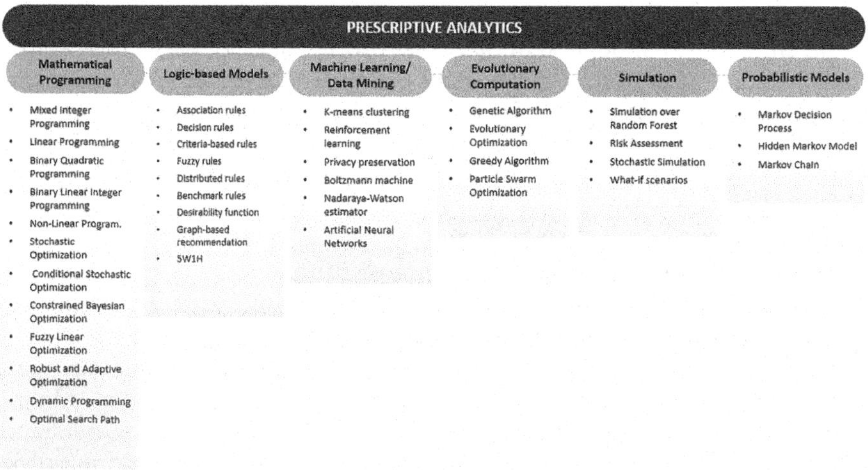

PRESCRIPTIVE ANALYTICS					
Mathematical Programming	**Logic-based Models**	**Machine Learning/ Data Mining**	**Evolutionary Computation**	**Simulation**	**Probabilistic Models**
• Mixed Integer Programming	• Association rules	• K-means clustering	• Genetic Algorithm	• Simulation over Random Forest	• Markov Decision Process
• Linear Programming	• Decision rules	• Reinforcement learning	• Evolutionary Optimization	• Risk Assessment	• Hidden Markov Model
• Binary Quadratic Programming	• Criteria-based rules	• Privacy preservation	• Greedy Algorithm	• Stochastic Simulation	• Markov Chain
• Binary Linear Integer Programming	• Fuzzy rules	• Boltzmann machine	• Particle Swarm Optimization	• What-if scenarios	
• Non-Linear Program.	• Distributed rules	• Nadaraya-Watson estimator			
• Stochastic Optimization	• Benchmark rules	• Artificial Neural Networks			
• Conditional Stochastic Optimization	• Desirability function				
• Constrained Bayesian Optimization	• Graph-based recommendation				
• Fuzzy Linear Optimization	• 5W1H				
• Robust and Adaptive Optimization					
• Dynamic Programming					
• Optimal Search Path					

FIGURE 9.8
Classification of prescriptive analysis techniques [26].

Digital Twin Models

The term "Twins" was used for the first time in NASA's Apollo program, where two or more identical versions of space vehicles were constructed, allowing the specialists on earth to use one of them to mirror the conditions of the space vehicle during the mission. Real-time data collected from the flight were used to represent the flight conditions and assist the astronauts in orbit in critical situations [29]. In 2010, the National Aeronautics and Space Administration introduced the term Digital Twin (DT) to describe models that presented three essential characteristics: robust simulation environment, high fidelity, and real-time connection with the physical version. Since then, they are increasingly being used in several areas of the industry 4.0, from original equipment in a process plant to an entire smart city that can be mirrored in a reliable virtual model. The models have been used for multiple purposes, such as optimization, virtual instrumentation (virtual reality assistant), and operational and equipment monitoring [30].

DTs models can be applied to create decision-support solutions or even be configured to adjust the process parameters to the optimal performance dynamically. Their models are intended to generate the virtual techniques for an object, being, or system in the digital approach to simulate their behaviors, using information from its physical units (such as sensor data) to make predictions, simulations, and dynamic change analysis. DT paves the way for the cyber-physical integration of process, being an essential factor for smart operations. Its operational principle combines simulation and optimization to guide the physical process to perform the optimal solution.

The concept of a DT is not new but empowered by an estimated 25 billion connected global devices by 2021 [31] through enhanced information technology and operational technology capabilities, and the last advances in BD and AI, DT models will be connected to the real world in real time for continuous monitoring and control, interacting with users do drive new business opportunities [32].

DTs are considered the next generation of simulation, and it is essential to understand the differences between both. The timeline in Figure 9.9 represents the most relevant changes in simulation systems until the current application of DTs.

The real-time aspect of DT is what most differentiates it from ordinary simulations, which are often used for offline optimization and design. A virtual model can be applied for the entire design–execute–change–decommission life cycle in real time. Highly connected via sensors, the real-world data is collated into the model, thus avoiding the problems associated with manual data collection, inconsistent updates, and difficulty of understanding the data. When physical things interact, standard simulations do not provide insights and intelligence that allow real-time intervention and long-term, continuous improvement. On the other hand, DTs are a source of comprehension that help operators incorporate this knowledge to make crucial decisions [33].

The capacity to collect data continuously and update a virtual version makes DTs a powerful tool for manufacturers and owners. In a mineral processing plant, DTs can replicate equipment, process, or the whole plant. At the heart of a DT is a model that mirrors the features and operation of the system [34]. For example, a DT model of a flotation cell can automatically adjust in operation, such as correcting the amount of reagent based on ore characteristics and other process parameters (bubble size, shape and color, float speed, and others). In this case, the decision-making rationale may be determined by experience and best practices (traditional automation setup) or, even better, optimized through the combination of predictive models and

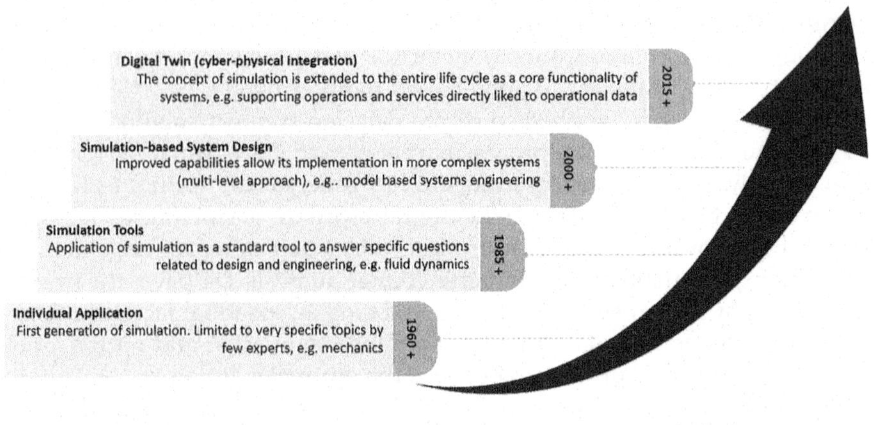

Digital Twin (cyber-physical integration)
The concept of simulation is extended to the entire life cycle as a core functionality of systems, e.g. supporting operations and services directly liked to operational data — 2015 +

Simulation-based System Design
Improved capabilities allow its implementation in more complex systems (multi-level approach), e.g.. model based systems engineering — 2000 +

Simulation Tools
Application of simulation as a standard tool to answer specific questions related to design and engineering, e.g. fluid dynamics — 1985 +

Individual Application
First generation of simulation. Limited to very specific topics by few experts, e.g. mechanics — 1960 +

FIGURE 9.9
Simulation evolution timeline [29].

AI algorithms or other sophisticated AAs methods (modern DT application). By combining current and historical data, the DT continuously forecasts the operational performance and equipment health and also predicts how a system can be automatically adjusted in critical situations and help uncover unexpected failures signs or operational issues before they occur by comparing historical, predicted, and actual data [23]. Those predictions can also be used in strategic decisions such as maintenance plans or definition of the best product mix. Moreover, some models have been used to test plant responses in abnormal conditions and for workforce training in a safe environment.

DTs are built as the optimal ultra-realistic version of a process or system, and the expected optimal behavior is compared with the current measurements to identify deviations and trigger actions in the control system towards the best performance.

Manufacturing is the most popular setting for DT. The biggest reason behind this is because manufacturers are always seeking for ways to track and monitor products to save time and costs [35]. DTs allow complete visualization of the manufacturing value chain, offering appropriate conditions for performing simulations and testing optimal parametrization in the entire system. In logistics, DTs have been applied for truck dispatch in Port and Mining operations. The model forecasts operational performance and calculates and assesses different dispatching alternatives. These models use real-time data extracted from its physical version to feed the DT model and to define the action triggers [36–38].

Augmented Analytics

Augmented Analytics develops potential data and information capabilities. It is transforming the lines between traditional Business Intelligence (BI) and AAs tools by enabling automation in all phases of the analytics cycle [39]. It is a promising solution to enhance the analytical process that facilitates the use of advanced ML and AI to extract value from data by experienced users [40]. Figure 9.10 presents some examples of AI applications through the set of activities of the analytics cycle.

Insights generated from earlier cycles in the analytics process can reveal new opportunities to redefine business problems. Automating the data preparation phase may dramatically save time and increase productivity. Data profiling and transformation work with data in different formats and magnitude. While data profiling phase identifies abnormal data distributions, null or inconsistent values, detect outliers, etc., data transformation suggests data cleaning, including treatment of null values, standardization, etc., reorganization including column splitting, aggregation, etc., blending and enrichment, including identification of join columns or suggestion of new data sets [40].

In the data discovery phase, analysis is enhanced by visualization tools such as Power BI, Tableau, Click View, and others. Those tools suggest different types of advanced data visualization based on the selected data,

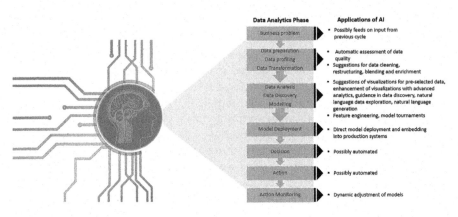

FIGURE 9.10
Applications of AI through the analytics cycle [40].

providing quick insights and levels of granularity analysis in a single interface. Furthermore, AAs can be applied to extract more in-depth insights; for example, from clustering analysis or forecasting models. The recent innovations in Natural Language Processing (NLP) have allowed data querying as an interactive feature for visualizations and automatic insights (trends, best performers, comparison with previous periods, key influencers, etc.) [40].

Platforms like Data Robot combine millions of features, ML algorithms, and model parameters to test and rank their performance. This method increases productivity and reduces the risk of biases [41].

In the deployment phase, model deployment can be integrated into production systems without extensive coding activities. The value created goes beyond model deployment and extends to model monitoring by incorporating new data automatically and retraining the model to optimize the performance of models already implemented by [40]. Normally, analytical tools do not consider decision-making and action taking in their scope. However, in the AAs world, operational decisions are increasingly being automated through ML models. Afterward, it may be necessary to take immediate action [40] automatically.

Based on the lack of data scientists, many companies are embedding augmented analytics in their data strategy. According to Gartner's prediction, Augmented Analytics will be responsible for automating more than 40% of data science responsibilities by 2020 [32].

All the AAs, as mentioned earlier, solutions applied for prediction, prescription, optimization, simulation, and, consequently, decision-making might significantly improve organizations' performance across various levels as well as guide their initiatives to manage critical global issues such as environmental and energy efficiency. The next section approaches an exceptional decision support solution that has been in the industry for decades and is undoubtedly taking advantage of emerging technologies and BD capabilities: Expert Systems.

Expert Systems

Expert systems (ES), sometimes referred to as Knowledge-based Systems [42], are a branch of AI that can solve high complex real-world problems using data and reasoning methods commonly associated with a human expert [43]. According to Turban and Aronson, they are the number one system in the significant group of solutions developed to incorporate knowledge in decision support systems, known as intelligent decision support systems or knowledge-based decision support systems [44].

They could also be defined as a computer system that performs at or near the level of an engineering specialist in a field of endeavor, providing intelligent decision-making support.

The basic principle behind ESs comprises the idea of transferring a human expert knowledge to a computer using programming techniques and advanced algorithms and consult the computer for specific advice. Similar to a human consultant (expert), it provides recommendations and the logic behind the advice [44]. Out of all current AI developments, ESs are considered the biggest one responsible for automating decision-making in engineering problem solving [45].

The application of ESs became very popular since its implementation in the 70s, especially in the 80s and 90s. Willian, 2017, conducted a rigorous analysis of 311 ES case studies published from 1984 through 2016, including a wide range of practical fields and problems domain [46]. The research provides an excellent overview of the presence and trends of ESs in the industry (Figure 9.11).

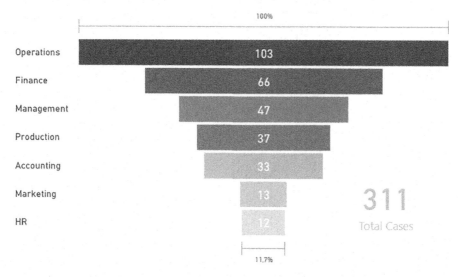

FIGURE 9.11
Distribution of case studies by functional area [43].

The wide range of applications across multiple industries is also demonstrated by the distribution of ES cases presented in Figure 9.12.

It is observed in Figure 9.12 that tech-savvy industries tend to invest more in ESs than other segments. Typically, they have a culture of experimentation, management of uncertainty, and are open to explore and adopt new technologies because of a higher potential payback [46].

Not differently from most of the tools, concepts, and methodologies discussed in this book, ESs have also evolved due to the great exposition to BD, IoT, increased computational power, and improved AI approaches. ESs have incorporated several AI techniques, such as control, data-driven, monitoring, and knowledge representation approaches [47]. Knowledge-processing techniques that are responsible for the organization and reasoning behind the ESs are mainly deployed for the following three objectives:

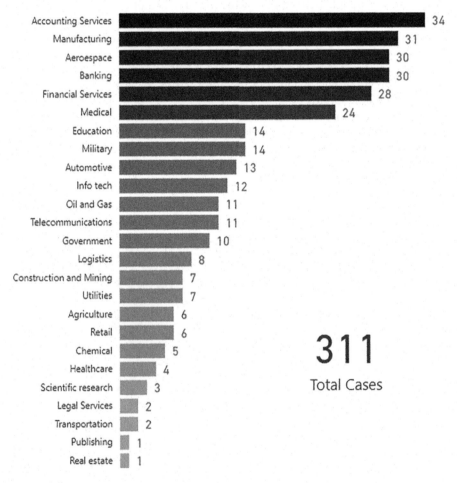

FIGURE 9.12
Distribution of ES cases by industry [43].

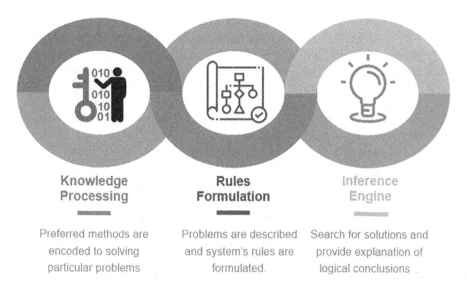

Knowledge Processing	Rules Formulation	Inference Engine
Preferred methods are encoded to solving particular problems	Problems are described and system's rules are formulated.	Search for solutions and provide explanation of logical conclusions .

FIGURE 9.13
Typical steps for knowledge extraction [47].

I. To enhance the reasoning performance of the application system

II. To extend the flexibility of the application system

III. To increase the level of human characteristics of the system [33]

The process of abstracting insights extracted from the knowledge processing can be described in three sequential phases (Figure 9.13).

ESs Components, Types, and Methodologies

ESs Components

Some core components form the architecture of ESs:

I. Knowledge Acquisition Subsystem

Knowledge acquisition subsystem is the part that contains all sources of knowledge incorporated into the system. Potential sources used to build or expand the knowledge base include human experts, textbooks, multiformat documents (audio, video, images), databases, high-relevance technical reports, and information extracted from the internet [44]. Its overall function is to provide appropriate conditions to capture and store information from the sources of knowledge in the knowledgebase [48].

II. Knowledgebase

The Knowledgebase contains all acquired knowledge sources required for problem description, comprehension, and resolution. In this context, knowledge can be separated into three types – compiled knowledge, qualitative knowledge, and quantitative knowledge. The knowledge-based component comprises two essential parts: First, facts, such as business problem and the correspondent hypotheses – described by an experienced employee – and second, rules or particular heuristics developed to solve the specified problem [44,48].

III. Inference Engine

The Inference Engine controls and interprets the rules described in the knowledgebase. It is considered the "brain" of the ES, playing the role of an advisor and consultant. Inference mechanisms are created to interact with the knowledgebase, searching for solutions. Essential functions to draw and explain logical conclusions are run in this component.

Complex decisions may demand a combination of multiple rules to represent expert knowledge and cover numerous conditions. The inference engine is responsible for dynamically chaining multiple rules together to conduct inference in an ES [44,48,49].

IV. User Interface

The user interface is the part of an ES developed to facilitate the interaction between the user and the system. In this module, the user accesses the recommended action and explanation given by the Inference Engine and, if necessary, ask questions to the system using a question-and-answer approach or another language processor [44,48].

V. Blackboard

Blackboard is an isolated working memory field destined for the explanation of a current problem as required by the input information. Hypotheses and decisions may also be recorded in this part of the system. In this context, decisions are classified into three types: a plan (how to attack the problem), an agenda (possible actions awaiting execution), and a solution (set of hypotheses and actions that the system has generated) [44].

VI. Explanation Facility (Justifier)

Explanation Subsystem is a crucial part of monitoring the responsibility for decisions to their sources during the knowledge transference phase and in problem-solving. Besides monitoring, when a user needs to know how the expert system arrived at a specific solution or question why some possible paths were excluded, the explanation facility can also explain the system's behavior through an interactive interface [44].

VII. Knowledge Redefining System

Like human experts, ESs can learn from their experience. In the Knowledge Refining System, causes for success and failures are analyzed to improve accuracy and practical reasoning [44].

All those seven different elements interact together in modeling the problem-solving process by an acknowledged human specialist of a specific field. Figure 9.14 shows the structural design of an ES with its elements and the way they cooperate.

The knowledgebase and the deference engine are the most critical components in an ES, as shown in the chart.

ESs Types

Based on the decision-making objective (business rule), different types of ESs can be applied. The most common types are as follows:

Forecast systems – Developed for making predictions, such as weather conditions, population growth, and agricultural production.

Troubleshooting type – This type relies on a robust database where it is possible to correlate parameters and make diagnoses. Maintenance and medical fields explore this type of system.

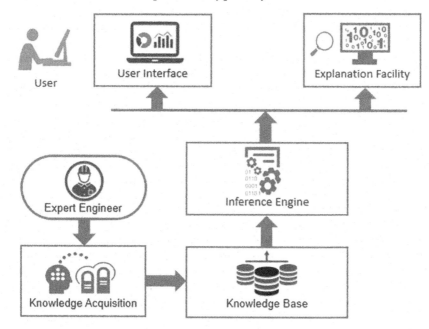

FIGURE 9.14
Architecture of an expert system.

Monitoring – Comprises systems with the ability to continuously track and compare performance metrics in some activities like marketing, transport, and production.

Educational expert system – It is formed by software packages created to teach students distinct principles of both physical and biological science.

Planning – In these systems, performance trends are compared with historical data and current demand to help planning activities such as annual budgets, allocation of funds, etc. [48].

ESs Methodologies and Techniques

Different authors describe ESs in terms of methodologies, techniques, and applications. The research conducted by Liao, 2005 [50], assessed 166 articles on ES applications and classified them into eleven categories according to their methodologies and applications. The number of classes tends to continually increase as more approaches are developed, including hybrid methods. The following methodologies are widely used in industrial applications.

Rule-Based Systems

In the Rule-base system, the information is presented in the form of a series of rules, such as IF-THEN, which combines the condition and the solution for dealing with a specific situation [50]. The IF parameter points to the condition for the rule to be checked, and the THEN parameter brings the action or conclusion if all IF conditions are satisfied [44]. Inferences can be performed applying the defined rules on data in order to find better solutions. Systems that use this methodology have high commercial applicability because of the robustness of advanced technology and the availability of development tools to end-users. This methodology is well suited for classification and diagnostic problems [45].

Knowledge-Based Systems

Also referred to as advisory systems, knowledge systems, intelligent job-aid systems, or operational systems, this type of system can efficiently perform activities that do not necessarily require an expert. The knowledge may be available in documented form or undocumented (expertise). A typical application is the automation of the help desk service using the knowledge-based methodology to analyze customers' profiles and recognize and address their needs [44].

Artificial Neural Networks

ANN systems are designed to simulate a biological neural network by developing artificial neurons. They can be implemented in ESs that depend upon many levels of logic without the engineering of a detailed knowledge structure. Differently of rule-based systems and other ESs, systems based on ANN store the knowledge implicitly. They are a black box that do not provide humans with an explanation about the logic or correlations in the data [49]. Systems are trained to learn the relation between input parameters and the output and then to evaluate the relation based on their relevance, thus developing theirs on solutions.

Fuzzy Expert Systems

These systems apply the fuzzy logic method, which deals with uncertainty. In this technique, computers are programmed to simulate human reasoning in less precise and logical behavior than performed by conventional computers. The principle behind this configuration is to reflect the uncertainty nature of the natural decision-making process [50,51].

Case-Based Reasoning

Case-based systems are built under the experience of previous cases solved (or not) by experienced human specialists. A robust database formed by past case's experiences is built for future retrieval when a new problem with similar characteristics is found. Database cases are compared with the current situation, and the solutions applied in the closest fit cases are identified and used for the solution generation process.

The new experiences, which include unsuccessful solutions and justification regarding the failures, are preserved and inserted in the database in the form of cases – rather than rules. This methodology is useful when the know-how of the problem is not fully available or difficult to obtain. Compared with other methodologies, it can provide solutions more quickly, avoiding the long time necessary to perform inferences typical of rule-based reasoning or the synthesis problem [45].

Finally, it is important to reinforce that an ES is not created to substitute the knowledge a worker must perform problem-solving tasks. However, these systems have an enormous potential to reduce the amount of work the individual must do to solve a problem, making a variety of complex decisions precisely, contributing to process stability, and leaving people with the creative and innovative aspects of problem-solving.

ESs in Mining

ESs have been used in a variety of applications in different phases of the mining value chain. They are a valuable tool to support decision-making in the following areas:

- Monitoring of environmental aspects of mining sites
- Analysis and forecasts of the mining deposits exploitation
- Performance management of the mining business
- Geomechanical evaluation and monitoring
- Geological modeling and feasibility evaluation
- Mining planning
- Forecast of gas dynamics for underground mines
- Operation management of open-pit mining
- Solving complex problems of mining rehabilitation [41,52]

Summary

AAs is a transformative tool to leverage the decision-making process. Most business leaders struggle in the face of the vast quantity and variety of data they need to deal with every day. Important decisions are taken based on intuition, and the potential of analytics is not fully explored to solve the organization's challenges. Many companies committed to evolving their digital maturity, embracing analytics and emergent technologies, and innovating problem-solving methods have been focusing on their organization structure redesign. An Agile structure has now come up as a more flexible and stable alternative, showing a new mindset that incentivizes rapid interaction and experimentation and promotes integration between the business and IT areas.

Furthermore, traditional performance management practices are being enhanced or replaced by AAs tools. Decision-support solutions have been dramatically improved by BD techniques and technologies currently available. Examples of those modern solutions are Digital Twin, Augmented Analytics, Action boards, Expert Systems, and others built under predictive and prescriptive models.

References

1. Zeleny, M., Strategy, and strategic action in the global era: Overcoming the knowing-doing gap. *International Journal of Technology Management*, 2008. 43(1–3): pp. 64–75.
2. PWC. Data-Driven: Big Decisions in the Intelligence Age. 2020 [cited 2020 28/01/2020]; available from: https://www.pwc.com/us/en/services/consulting/library/view-of-decisions.html.

3. Smet, A., G. Jost, and L. Weiss, Three Keys to Faster, Better Decisions. 2019 [cited 2020 28/01/2020]; available from: https://www.mckinsey.com/business-functions/organization/our-insights/three-keys-to-faster-better-decisions.

4. Smith, T., et al., Analytics and AI-Driven Enterprises Thrive in the Age of With. 2019 [cited 2020 28/01/2020]; available from: https://www2.deloitte.com/content/dam/Deloitte/ec/Documents/technology-media-telecommunications/DI_Becoming-an-Insight-Driven-organization%20(2).pdf.

5. Elgendy, N. and A. Elragal, Big data analytics in support of the decision making process. *Procedia Computer Science*, 2016. 100: pp. 1071–1084.

6. Singh, H., Using Analytics for Better Decision-Making. 2018 [cited 2020 25/01/2020]; available from: https://towardsdatascience.com/using-analytics-for-better-decision-making-ce4f92c4a025.

7. Aronowitz, S., A. De Smet, and D. McGinty, Getting organizational redesign right. *McKinsey Quarterly*, 2015. 9: pp. 1–11.

8. Tollman, P., A. Toma, and F. Roghé, A New Approach to Organization Design. 2016 [cited 2020 28/01/2020]; available from: https://www.bcg.com/en-br/publications/2016/people-organization-new-approach-organization-design.aspx.

9. Cardoso, D.M., Taylor's Scientific Management Principles in Current Organizational Management Practices. 2014 [cited 2020 25/01/2020]; available from: https://www.academia.edu/9553271/Taylors_Scientific_Management_Principles_in_Current_Organizational_Management_Practices.

10. Aghina, W., K. Ahlback, and A. Smet, The Five Trademarks of Agile Organizations. 2018 [cited 2020 25/01/2020]; available from: https://www.mckinsey.com/business-functions/organization/our-insights/the-five-trademarks-of-agile-organizations.

11. McKinsey. McKinsey Organization Agility and Organization Design. 2016 [cited 2020 28/01/2020]; available from: https://www.mckinsey.com/~/media/McKinsey/Business%20Functions/Organization/Our%20Insights/McKinsey%20on%20Organization/McKinsey%20on%20Organization%20Agility%20and%20organization%20design.ashx.

12. Porsche Consulting, Agile in a Complex World How Organizations Become Both Flexible and Stable. 2016 [cited 2020 28/01/2020]; available from: https://www.porsche-consulting.com/fileadmin/docs/04_Medien/Publikationen/SRX03904_Agile_in_a_Complex_World/Agile_in_a_Complex_World_EN_Porsche_Consulting_v5.pdf.

13. Denning, S., How to make the whole organization "Agile". *Strategy & Leadership*, 2016. 44: pp. 10–17.

14. Zakaria, Z., A cultural approach of embedding KPIs into organisational practices. *International Journal of Productivity and Performance Management*, 2015. 67: pp. 932–946.

15. Maté, A., J. Trujillo, and J. Mylopoulos, Specification and derivation of key performance indicators for business analytics: A semantic approach. *Data & Knowledge Engineering*, 2017. 108: pp. 30–49.

16. Peral, J., A. Maté, and M. Marco, Application of data mining techniques to identify relevant key performance indicators. *Computer Standards & Interfaces*, 2017. 54: pp. 76–85.

17. Black, Q., Predictive Maintenance, Improving Early Failure Detection and Reducing Machine Downtime. 2019 [cited 2020 28/01/2020]; available from: https://quantumblack.com/work/predictive-maintenance/.
18. LaValle, S., et al., Big Data, Analytics and the Path from Insights to Value. 2010; available from: https://sloanreview.mit.edu/article/big-data-analytics-and-the-path-from-insights-to-value/.
19. Soofastaei, A., Development of an advanced data analytics model to improve the energy efficiency of haul trucks in surface mines. In School of Mechanical and Mining Engineering. 2016: The University of Queensland, School of Mechanical and Mining Engineering, Brisbane, Australia.
20. Kota, L., Artificial intelligence in logistics. *Advanced Logistic Systems-Theory and Practice*, 2018. 12(1): pp. 47–60.
21. Zhang, Y., The application of artificial intelligence in logistics and express delivery. *Journal of Physics: Conference Series*, 2019. 1325, p. 021085. IOP Publishing.
22. Wang, H., et al., Towards felicitous decision making: An overview on challenges and trends of Big Data. *Information Sciences*, 2016. 367: pp. 747–765.
23. Glaessgen, E. and D. Stargel, The digital twin paradigm for future NASA and US Air Force vehicles. In 53rd AIAA/ASME/ASCE/AHS/ASC Structures, Structural Dynamics and Materials Conference, pp. 23–26. April 2012 Honolulu, Hawaii.
24. Brink, H., M. Prema, and K. Reisman, Intelligent Actionboards: Stop Staring at Dashboards and Start Getting Things Done. 2010 [cited 2020 28/01/2020]; available from: https://www.mckinsey.com/~/media/McKinsey/Business%20Functions/Operations/Our%20Insights/Intelligent%20Actionboards%20Stop%20staring%20at%20dashboards%20and%20start%20getting%20things%20done/20120822_intelligent_actionboards.ashx.
25. Kelleher, J.D., B. Mac Namee, and A. D'arcy, *Fundamentals of Machine Learning for Predictive Data Analytics: Algorithms, Worked Examples, and Case Studies*. 2015: MIT Press, Cambridge, MA.
26. Lepenioti, K., et al., Prescriptive analytics: Literature review and research challenges. *International Journal of Information Management*, 2020. 50: pp. 57–70.
27. Gartner. Prescriptive Analytics. 2019 [cited 2020 28/01/2020]; available from: https://www.gartner.com/en/information-technology/glossary/prescriptive-analytics.
28. Soofastaei, A., Introductory Chapter: Advanced analytics and artificial intelligence applications. In Soofastaei, A. (ed.) *Advanced Analytics and Artificial Intelligence Applications*. 2019: InTechOpen, London. pp. 10–22.
29. Boschert, S. and R. Rosen, Digital twin—the simulation aspect. In Hehenberger, P., Bradley, D. (eds.) *Mechatronic Futures*. 2016. Springer, Cham. pp. 59–74.
30. Kortelainen, J., Utilization of digital twin in model-based control of flotation cells. 2019; available from: https://lutpub.lut.fi/handle/10024/160275
31. Purdy, M., et al., How Digital Twins Are Reinventing Innovation. 2020 [cited 2020 30/01/2020]; available from: https://sloanreview.mit.edu/article/how-digital-twins-are-reinventing-innovation/.
32. Panetta, K., Gartner top 10 strategic technology trends for 2019. Published online October 15, 2018.
33. Raghunathan, V., Digital Twins vs Simulation: Three Key Differences. 2019 [cited 2020 30/01/2020]; available from: https://www.entrepreneur.com/article/333645.

34. Jämsä-Jounela, S.-L., Future automation systems in context of process systems and minerals engineering. *IFAC-PapersOnLine*, 2019. 52(25): pp. 403–408.

35. Fuller, A., Z. Fan, and C. Day, Digital twin: Enabling technology, challenges and open research. arXiv preprint arXiv:1911.01276, 2019.

36. CoalAge. Miners Turn to Digital Twins for Asset Performance Gains. 2018; available from: https://www.coalage.com/features/miners-turn-to-digital-twins-for-asset-performance-gains/.

37. Hofmann, W. and F. Branding, Implementation of an IoT-and cloud-based digital twin for real-time decision support in port operations. *IFAC-PapersOnLine*, 2019. 52(13): pp. 2104–2109.

38. Collins, B., Anglo Using 'Digital Twins', Robotics to Boost Mining: Q&A. 2018; available from: https://about.bnef.com/blog/anglo-using-digital-twins-robotics-boost-mining-qa/.

39. Sallam, R., Just Buying Into Modern BI and Analytics? Get Ready for Augmented Analytics, the Next Wave of Market Disruption. 2017 [cited 2020 25/01/2020]; available from: https://blogs.gartner.com/rita-sallam/2017/07/31/just-buying-into-modern-bi-and-analytics-get-ready-for-augmented-analytics-the-next-wave-of-market-disruption/.

40. Prat, N., Augmented analytics. *Business & Information Systems Engineering*, 2019. 61(3): pp. 375–380.

41. Knight, W., You Could Become an AI Master Before You Know It. Here's How. 2017 [cited 2020 30/01/2020]; available from: https://www.technologyreview.com/s/608921/you-could-become-an-ai-master-before-you-know-it-heres-how/.

42. Shiue, W., S.-T. Li, and K.-J. Chen, A frame knowledge system for managing financial decision knowledge. *Expert Systems with Applications*, 2008. 35(3): pp. 1068–1079.

43. Leo Kumar, S., Knowledge-based expert system in manufacturing planning: state-of-the-art review. *International Journal of Production Research*, 2019. 57(15–16): pp. 4766–4790.

44. Aronson, J.E., T.-P. Liang, and R.V. MacCarthy, *Decision Support Systems and Intelligent Systems*. Vol. 4. 2005: Pearson Prentice-Hall, Upper Saddle River, NJ.

45. Rajeev, S. and C. Krishnamoorthy, *Artificial Intelligence and Expert Systems for Engineers*. 1996: CRC Press, Boca Raton, FL.

46. Wagner, W.P., Trends in expert system development: A longitudinal content analysis of over thirty years of expert system case studies. *Expert Systems with Applications*, 2017. 76: pp. 85–96.

47. Cortez, P., et al., Insights from a text mining survey on expert systems research from 2000 to 2016. *Expert Systems*, 2018. 35(3): p. e12280.

48. Ravindranath, B., *Decision Support Systems and Data Warehouses*. 2003: New Age International, New Delhi.

49. Looney, C.G., Neural networks as expert systems. *Expert Systems with Applications*, 1993. 6(2): pp. 129–136.

50. Liao, S.-H., Expert system methodologies and applications—a decade review from 1995 to 2004. *Expert Systems with Applications*, 2005. 28(1): pp. 93–103.

51. Mohammed, A.A., et al., Expert system in engineering transportation: A review. *Journal of Engineering Science and Technology*, 2019. 14(1): pp. 229–252.

52. Chen, J., et al., An expert system for metal resources exploration and mining feasibility evaluation. *Geo-Resources Environment and Engineering (GREE)*, 2017. 2: pp. 72–77.

10

Future Skills Requirements

Paulo Martins and Ali Soofastaei

Advanced-Data Analytics Company Profile – Operating Model

The Analytics of big data has been one of the main topics in engineering for many years, and no one can deny that the application of Advanced Analytics (AA) tools is transforming the way many companies do business and opening opportunities to competitive advantage. There are some related questions in this field. For example,

- what are the formula or critical success factors for exploiting data analytics? and
- how can a company be transformed to take advantage of data and advanced analytics tools?

Behind all efforts for successfully deploying AA, Artificial Intelligence (AI), Machine Learning (ML), Deep Learning (DL), companies need to rethink their data culture and focus their strategies to emphasize data and analytics. Recently completed research by McKinsey shows that there is a gap between laggards and leaders in adopting AA. These days, we see some companies are doing wonderful tasks; many are still trying with the basics; and many are feeling undeniably overwhelmed, with managers and members of the rank and file questioning the return on information initiatives [1].

To achieve this goal, many companies have established hybrid organizations, which include centers of excellence, analytics sandboxes, or innovation labs to derive benefits more rapidly from their data investments [2]. For example, BHP set up its Innovation Center in 2017 to develop products and services that integrate data and devices to make BHP's work more accurate than ever [3]. Another mining company that has intensified the data analytics investment is Vale, which inaugurated the AI Centre in 2019 as part of the evolution strategy that aims to influence the implementation of innovative and disruptive technologies in all sections of the mining business, always focusing on the results through data analysis and process improvements [4].

Further examples of mining companies leading the data analytics race are Rio Tinto [5], Barrick Gold [6], and Anglo American [7].

In a digital analytics strategy, mining companies can explore three sources of value levers: value creation (encompass projects that improve production or product quality), value protection (encompass projects enhance safety or deal with environment or community challenges), or value definition (projects that involve the definition of ore reserves) [8].

What Is and How to Become a Data-Driven Company?

Data Analytics is one of the buzz words nowadays. Becoming data-driven has been a top priority for many companies, but not all fully understand the benefits of analytics and are unsure about how to proceed. What exactly does this mean? It is essential to put data and analytics at the core of the business strategy, its structures, processes, and the culture rather than simply installing the right tools and applications. All sources of data are treated as valuable resources, and employees are committed to using the right data at the right time to foster better decisions [9].

Some factors have a decisive influence in the current context to favor the implementation of AA. The volume of available data has increased, accessible in (near) real time, both from internal and external sources. Storage capacity, processing power, and available bandwidth have continued to increase exponentially at decreasing costs. The development of new data structures, collaborative environment, and easy-to-integrate services have opened the avenue to new, practical applications.

Despite all factors, as mentioned earlier, to explore the values that can be captured through Advanced Data Analytics, companies should start their transformation journey by overcoming typical roadblocks, such as corporative culture, talent acquisition, and retention and technology. Some businesses have supported technology heavily as the very first step to data-oriented growth, but this alone is not enough. Companies have to take the human side of data much more seriously and creatively if they genuinely expect significant business benefits [2].

Corporative Culture

A survey published in 2019 by New Advantage Partners reported that 72% of the interviewed executives had not yet forged a data culture. The authors have focused on the role of companies in shifting human attitudes and behavior around data rather than investing the vast majority on big data and information technology (IT) in technology and growth [10].

The human side of this cultural change starts from leadership understanding about the difference between AA and traditional analytics. Success with analytics requires endorsement from senior leaders by explicit support throughout all levels of the organization. A change in management approach is required, and it includes a shared mindset that recognizes data and insights as the catalyst towards a more objective and efficient decision-making process followed by an extensive communication effort. A collaborative environment must be maintained to connect data and processes in order to make data and lessons learned affordable for all departments. As data science is rapidly evolving, continued education on the latest approaches and techniques and benefits they can provide is required to promote the organization's culture and empower decision-makers to deal with more complex questions. Hiring data scientists is a need to leverage the application of sophisticated analytics models and identify improvement levers. Companies can also deploy data translators (analytics translators) who are trained to build the bridge between the operators, IT specialists, and analytics experts. Moreover, instead of undertaking massive change, companies should focus their actions on the source of data, build models, and transform the organizational culture. When faced with limited resources, such as IT infrastructure and small team, a more gradual approach is recommended, avoiding high complexity by developing a pilot case instead.

Talent Acquisition and Retention

It is intuitive that organizations who struggle to develop a data culture also have limitations to develop their workforce. There is a secure connection between both, and data-driven organizations employees are better educated on data concepts. However, it is a significant challenge for businesses in all sectors to attract and retain the best talent, and most organizations have little capacity to recognize problem areas and potential solutions for recruitment challenges. Now more than ever, harnessing AA opportunities involves various disciplines, and some roles such as data engineers, data scientists, data architects, and data translators, are new to most traditional companies, and related competencies are rarely found in-house. External hiring of new talents is mandatory and should go hand in hand with an upgrade of the existing workforce's skills.

Companies must look into and comment on the whole employee experience in order to win over the competition and attract good people. A distinct culture, career paths, and data and analysis talent strategy must be identified, the link between employees and business leaders strengthened, and the specific contributions made by data and analytical talents must be articulated. New or alternative talent sources must be established and explored – retraining existing staff or establishing creative, strategic collaborations, for instance [11].

Massive demand for data scientists and Big Data analysts has been created in the market. Data Scientists have consecutively been appointed – from 2016 to 2019 – and are the number one top position in America's Top Job List according to the employment site Glassdoor [12]. The 2019 list also includes Data Engineer, Business Analyst, Solutions Architect, Systems Engineer, and Data Analyst. All those roles have growth forecasts, which indicate high competition for talent acquisition.

Technology

To support a data analytics strategy, companies need to ensure that the underlying technology is available by deploying a modern data architecture. With a robust and modern data architecture in place, organizations can support quick data collection and data sharing that allows lead workers to access and use the data they require. Furthermore, it helps to establish and maintain the high levels of data quality necessary to help effective data-based decision-making [13]. Most of the traditional legacy systems were not created to deliver a continuous flow of information in real time, and business leaders need to work closely with chief information officers to define and prioritize the most relevant data for the application of AA. Additionally, cloud storage is also a valuable alternative to scale computing power to embed big-data demands – higher speed and flexibility – cost-effectively [14]. The overall trend across industries is a migration of excellent IT infrastructure to the cloud.

The competitive advantage comes from establishing which data are relevant to address business problems and accessing a significant amount of consistent data, implementing high-velocity data experimentation, automated workflows along the data life cycle, and new ways of thinking about business challenges. Only a fraction of the overall value creation comes from algorithms, which is also reflected in the difficulty companies have in protecting their algorithms, since most of them are open source.

The Profile of a Data-Driven Mining Company

As the mining industry has to meet new challenges, such as declining resources access and quality, sudden fluctuations in commodity prices, tightening resource nationalism and regulation, and increasing customers' requirements, the future mining engineers should prepare themselves for firmer operating conditions. It will need more insights from data and better collaboration with suppliers, communities, employees, and customers to be successful.

Mining decisions will rely on real-time data extracted from processing equipment and sensors during the operation. Accessing this data across the mining value chain will enable miners to identify critical drivers of process variability and drive rapid operational improvements. Additionally, there will be a possibility to update their ore bodies, mine plans, and financial models more frequently while shortening planning cycles [15].

In many cases, companies start with improved visualization tolls, integrating data from multiples sources and decreasing reliance on disconnected systems to facilitate improved analysis of information. Success also involves the development of data science and analytical skills throughout the organization to turn growing volumes of data into meaningful insight. Some successful cases of AA implementation are described in one of the recently published articles by Mckinsey titled "Behind the mining productivity upswing: Technology-enabled transformation" [16].

In some mines, AA and ML systems collect all the data and use it to enhance procedures in the concentration area. This approach is optimal globally rather than optimal locally via automated process control. This can increase energy efficiency. It is now not unusual for mineral recovery to be increased by 1%–3%, and the output to be enhanced by 4%–8% while energy consumption decreases. It means an increase in productivity of 5%–10%, which almost means opening a new mine when used over the footprint of a regular mining company, not including the capital cost [16].

One metal mine uses ML and AA to construct an industrial device that increases performance and improves mineral quality. The business collects information for small to large plants from more than 100 sensors. An algorithm forecasts the plant performance and offers a range of proposals based on achievable parameters to optimize it. The plant managers who make the decisions are reviewing the program. The effect of the applied modifications is continually calculated to refine the system [16].

Another example in this area shows how a metal mine used internet of things (IoT) sensors, combined with a central dataset and advanced ML, to boost the chemical recuperation from mining by 10%–15%. The real-time data gathered from the ML model calculates the ideal variables for sodium hydroxide recovery that improved performance over that previously achieved. This has saved millions of dollars in chemical capital and the effect of pollution on the environment [16].

The following sections will explain the insights and perspectives about the transformative impact AA and other emerging technologies will have on the future of the mining workforce [17].

Jobs of the Future in Mining

The last section briefly presented some considerations about the skills shortage faced by companies that are looking for embedding AA in their corporative culture.

The introduction of new technology redevelops current positions across the mine value chain, and today we do not know many of the tasks, skills, and job titles. Smart systems, automation, and AI are advancing quickly, reshaping the workforce of the future and adjusting the abilities organizations are looking for in their individuals.

Technology has the power to significantly improve our lives, enhancing efficiency, living requirements, and median life span, freeing individuals to place on emphasis on personal fulfillment [18]. All those possibilities raise fundamental questions:

- Who will be running the mine of the future, and what skills will be needed?
- How can mining companies create an attractive workplace to win the competition for young talents?
- What changes will be necessary for the educational agenda to efficiently adapt and overcome the skills shortage in the mining industry?

The high volume of data generated by a digital mine needs to be appropriately managed to deliver real-time insights, and talents from non-traditional resources can offer a variety of abilities and the mindset that will prepare miners to overcome future challenges. The new frontier of mining jobs will undoubtedly include areas such as AA, AI, ML, automation, robotics, and IoT. The more mining companies invest in innovation and technology, the higher is the impact on the work, workers, and mining workplace, for both front-line employees and management. Introducing the latest applications and recruiting people from the gaming or "joystick" generation will not alone guarantee a company's analytics success. It is essential to place digital tools at the center of an operational mindset and fully embed them in the total flow of the work [19]. To manage all these innovative methods and tools, organizations need to be ready to adapt quickly and create a training agenda to upskill the new generation of employees.

In the mine of the future, most of the manual and routine tasks that require minimum cognitive effort will be replaced by Robotic Process Automation and cognitive assistants. However, instead of replacing humans, technology will change the modality of work and, in some cases, create new jobs to integrate them in order to prosper in this new changing landscape. People will be retrained to use technology, and jobs will be redesigned to take better advantage of existing social skills [15].

The natural trend is to separate workers and break the roles so that businesses can determine the best resources to execute each task, be it conventional, outsourced, or digital technology employees. The ideal end stage is to automate tasks that do not add value and allow the right person to complete or question tasks of interest. Once every job is so coordinated, mining companies will start to build a vision for their future employees – the roadmap of skills they believe will be the most critical demand. It helps them to master their future – developing strategies to fill their talent holes before the competitors even know which gaps exist [20].

Some fields like computer technology, IT, statistics, mathematics, and data science will be essential to manage this transition to the digital mine since they help to develop critical and analytical thinking, interpretation of data, and problem-solving skills.

Activities that were executed in groups may come to be completed by individual employees supported by various types of machines that provide digital assistance [21]. Connected solutions that focus on the integration from mine to port are developed to process online information from personnel and machinery and manage the complete operation from a control room (integrated remote operating centers), sometimes hundreds or thousands of kilometers away from real operation. Driverless trucks and trains, automated drills, and other automation initiatives are entirely changing the way mining companies allocate staff, minimizing risk and operational costs while increasing efficiency and productivity.

Real-time analysis supports decision-making using data collected from thousands of sensors, cameras, and imaging technologies. These new devices generate vast amounts of geospatial, equipment health, and operational data. For instance, a Rio Tinto mining truck with 220 sensors generates about five terabytes of data per day, and the entire Rio Tinto iron ore business generates approximately 3.5 petabytes of data per day [22]. BHP is using a downhole tool to collect and log elemental distributions for real-time analysis. Vale has improved its asset management supported by the application of AI models in data collected from mine equipment, such as off-road trucks, excavators, and loaders [23,24]. All those examples reinforce the trend of increasing demand for Data and Data Literacy skills, where professionals have a complete domain of digital technologies and data analytics tools and methods.

In a detailed analysis of the future of the workforce in the mining industry, the consulting firm Ernest Young defined a methodology to quantify the impact of technology, dividing the mining value chain into six stages and considering the diversity of occupations in each stage [25]. The approach created a probabilistic model called the Technology Impact Index, which provides the following segmentation to classify different levels of impact:

- **Enhanced (probabilities between 0 and 0.3)** – Indicates that these occupations will be enhanced by technology and digital assets, and a few changes will be necessary to adjust for the future.
- **Redesigned (probabilities between 0.3 and 0.84)** – Indicates that some aspects will need to be redesigned, while some functions will be replaced and some other upskilled.
- **Automated (probabilities between 0.85 and 1)** – It is the most likely automation classification, indicating that the majority of tasks will be replaced, causing an occupation reduction.

Figure 10.1 presents the results for the six stages of the mining value chain.

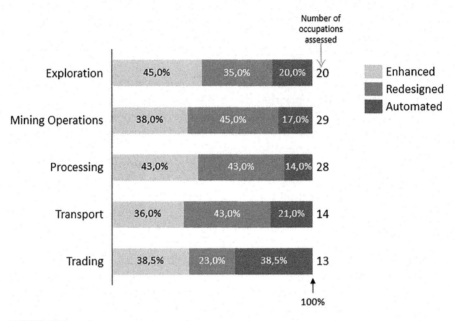

FIGURE 10.1
Occupation type by value chain [25].

In the Exploration phase, there is expected to be a reduction in drill operators, traditional surveyors, and field geologists, while there is an increasing demand on data analytics and modeling skills and better communication with centralized decision-making management [26]. When it comes to Mining Operations, results reveal the most considerable proportion in terms of redesign caused by the optimization and integration with technology, such as drones, integrated systems, and geospatial technologies, which will lead to fundamental changes in operations and workforce. Decision support will be the focus rather than the execution, and the relationship between people and assets will be reshaped. Consequently, there will be a reduction in drill operators, traditional mine geologists, truck operators, and on-site employees, while the market for data analytics and design skills will grow. Data scientists, data translators, geotechnical engineers, and modelers will be in high demand. The processing stage will be impacted by a fully integrated mine-to-plant approach trend. Digital tools will support the integration between upstream and downstream operations. Characterization and data from the upstream process will be utilized for predictive analysis and optimized decisions. Data, analytics skills, and decision support for plant setpoint management will increase.

The future of transport will see a reduction of the on-site workforce, and significant enhancements and redesigns in its occupations due to the implementation of autonomous assets (rails, conveyors systems, etc.) and optimized and integrated planning have been enabled by technologies, such as AA, ML, automation, digital twin, and others. Trading is the phase that has a more significant number of occupations susceptible to automation. It will require marketing professionals with an integrated view and knowledge of product development. A customer-centric approach and its integration upstream into the supply chain is the new focus leveraged by solutions, such as blockchain, market forecasting, and modeling methods. The end-to-end value chain shows the highest quantity of occupations enhanced by technology. A never-before-seen integration across the mining value chain will improve management and optimization. The driver to this transformation is a greater integration of roles and responsibilities combined with an operating model, performance metrics, and target setting alignment. The expected changes will be supported by the implementation of AA, digital twin, and integrated technologies.

According to the International Institute for Sustainable Development, as a result of the successful implementation of autonomous haul trucks, the operational workforce will be profoundly impacted in the short term. In the study, advanced control systems will have a long-term effect too, but it is still contingent on other automation technologies being implemented [27]. Although automation will impact some occupations, it is demonstrated that the enhancements and productivity benefits delivered by technology will be far more significant. The impact of technology in the current occupations was assessed to determine the occupation mix in the future. Results for those that represent more than 1% of the current workforce and the workforce of the future are presented in Figure 10.2. For visualization purposes, data were divided into two groups: high proportion group – occupations which currently represent more than 3% of the total workforce – and low proportion group – those occupations which represent less than 2% of the total workforce.

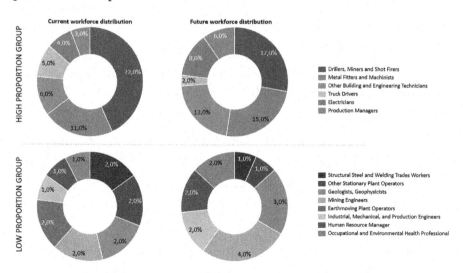

FIGURE 10.2
Proportion of current and future workforce [25].

Figure 10.2 confirms the tendency of reduction in the number of operational jobs associated with manual and routine tasks, as shown by the differences between current and future workforce in the groups that involve Plant Operators (stationary and earthmoving), Truck Drivers, Drillers, Miners, and Shot Firers. New requirements for roles in the creation, monitoring, management, and maintenance of autonomous remotely controlled equipment and data analysis and systems will be developed [27]. Consequently, the mine of the future will see a more technologically knowledgeable workforce that combines technical mining skills with digital technological competency. At the same time, newer functions such as data sciences, data engineers, data architects, data translators, modelers, etc. will provide core practical support [25].

Future Skills Needed

The speed of technological disruption and evolution are demanding more high-level responsiveness and innovation from organizations to retain their competitive advantage. Intelligent mines need specific skill sets, and understanding the current gap is crucial to prepare companies to plan their workforce and sustain competitive advantage strategically. As discussed in the last section, the Ernest Young report predicts that the request for digital literacy skills and information across all phases of the mining value chain will keep growing, leading to a redefinition of most of the functions linked with human-to-machine interaction. The incorporation of these skills will play an essential role in the decision-making process, optimizing everyday work [25].

The document points out that emerging technology, such as helicopters, autonomous vehicles, and remote-controlled operating systems, would enhance mining operation and exploration. Integrated operating centers allows more work to be performed remotely and flexibly, keeping employees away from the risks on-site. Traditional mining jobs such as truck drivers, drill operators, supervisors, and field geologists will be redefined based on the data and digital skills required to manage emerging technologies. The report also shows that the future distribution of occupation types will also impact skills demand. Figure 10.3 lists the skills with growing demand and those that will be reduced.

The analysis conducted in the report identified that the following skills would be of higher relevance for the occupations in the mining value chain and also the most resistant to the impact of technology:

- Creativity
- Change Management
- Design Thinking

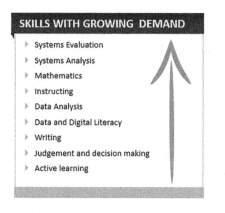
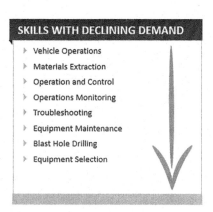

SKILLS WITH GROWING DEMAND

▸ Systems Evaluation
▸ Systems Analysis
▸ Mathematics
▸ Instructing
▸ Data Analysis
▸ Data and Digital Literacy
▸ Writing
▸ Judgement and decision making
▸ Active learning

SKILLS WITH DECLINING DEMAND

▸ Vehicle Operations
▸ Materials Extraction
▸ Operation and Control
▸ Operations Monitoring
▸ Troubleshooting
▸ Equipment Maintenance
▸ Blast Hole Drilling
▸ Equipment Selection

FIGURE 10.3
Movement of skills [25].

- Data Analysis
- Data and Digital Literacy
- Stakeholder Analysis
- Complex Stakeholder Engagement
- Strategic Planning
- Collaboration

A more detailed representation focused on the operator of the futures was developed by Low, Abrahamsson, and Johanson [21], who modified the typology of the future Industry 4.0 operator formed by the Romero et al. [26] to relate the eight worker stereotypes to the future mineworkers.

- The super-strong miner uses biomechanical support to boost the mobility and power of his limbs.
- The advanced miner uses increased reality to incorporate data from the digital to the physical world. Models have maintainers who are funded directly by equipment manufacturers. In individual glasses receiving and sending live videos, the problems could be solved by both partners via the manufacturer's guidelines.
- The virtual miners use virtual reality to simulate dangerous real-life situations and aid in training. Virtual reality training is relatively common in mining [28] and is perhaps one of the industries where it is the most useful. This includes planning for high-risk cases, such as fire and the introduction of new models. In principle, it is possible to put the entire management of production and maintenance in a virtual situation and thus make it unbiased.
- For health measurements as well as GPS position, the healthy mineworker uses wearable sensors. In particular, these reforms have

also been made in the mining industry. As an example, improved technologies are used, and projects are being carried out to incorporate sensors to monitor the health of miners.

- The cleverer mineworkers use smart helpers to connect databases, computers, machines, and other IT systems. There are many examples at present where radiofrequency identification tag systems and an intelligent application are used for prompt and multifunctional devices.

- For Very different, challenging jobs, the cooperative mine worker uses joint robots. There are semi-independent systems on the sites such as loaders and excavators where the operator "structures" the unit first and then automatically controls it. At the same time, the worker remotely directs it for the duration of the actual loading. The repetitive and frustrating task of driving from and to the muck is done by the vehicle.

- For contact between employees, businesses, and IoT, social networking workers make use of company social networking. This has been made possible by the development of public Wi-Fi and 5G. Although most mines do not have dedicated corporate social networks, Internet access in mines often means social media access. In the past, there were mines in which underground workers used group chats to exchange information by using mine-spreading Wi-Fi.

- The miner uses big data analysis to discover useful information and to forecast specific occurrences.

The researchers named this new operator profile as "Miner 4.0" and concluded that the possibilities created by technology would influence the future workforce. On the other hand, mining companies need to be prepared to respond to all those changes correctly. According to the Deloitte's Future of Work report, the subject is not about frontier employees being able to adapt and preserve pace with disruption and adjustments to work, but around the organization's ability to adapt along with them [19].

Challenges

Many of the critical resources mining companies need for the future will not be available in-house, and developing the required capabilities in-house can take years. Developing, hiring, and retaining the appropriate talent would require a sustained effort and a robust strategy. Moreover, besides mapping future roles, mining companies should work with research centers, mining schools, and universities to ensure the scholars are trained with the skills needed to work in the mines of the future. The rate of change promoted by technology is challenging the capacity to attract, develop, and retain the talent required to run the mine of the future.

Need for Mining Engineering Academic Curriculum Review

The last developments driven by technology are motivating a complete review in the roles and the competences of the mining engineer. Additionally, universities are facing a decline in the number of engineering and mining-related degrees [29,30], and there is a gap between the skills required to run a digital mine and academic curriculum learned in the institutions. The current scenario requires a more focused and specialized skill set. Modern mining companies are capital- and technology-intensive, with complex processes and industrial systems [31. Prerequisites to succeed in the digital world and transform the next-generation mine include the capacity to enable technologies and systems engineering, change management, and innovation [32].

According to the Global Mineral Engineering Curriculum Review developed by the IMPC Commission on Education, there is a difference between what is being taught in some minerals engineering courses and what the current industry needs [33]. The industry that undergraduate students will face has dramatically changed compared to the one for which much of the mining curriculum was created [33]. In the SAP's Future of Mining report, Finlayson stated: "If you take a mining engineer curriculum when Curtin (University) took over WAIT (Western Australian Institute of Technology) 50 years ago compared to now, it is the same curriculum bar one unit." [34].

The focus of mining engineering curricula in Australia has always been around academic requirements associated with mine manager's statutory certificate of competency. Although those skills are still essential, the future mining engineer needs to be able to absorb the dynamic of technology advances as well as address global challenges such as the paradigm of sustainable mining [32]. According to predictions by different researches, 80% of the current technologies will be out of date in the next ten years, and 80% of the workforce will have an education acquired more than ten years before that [31]. Therefore, education providers need to redesign their strategy based on a deep understanding of technological trends and changing practices in order to build an appropriate workforce to run the mine of the future [33]. The Mines of the Future report highlighted that educational changes should encompass not only courses and content but also other facets such as delivery mode and menu of educational offerings [8].

Fortunately, the standard view recognizes the urgent need to modernize the academic curriculum in order to meet digital age challenges, including teaching with digital devices such as IoT, blockchain, and AI. Universities are using several internet webinars and courses, such as mining education in Australia, University of British Columbia mining, and Edu-Mine, to complement their curricula as a valuable source of high-quality, modern studies [35].

A team of academic staff and industry experts at the University of the Witwatersrand in the School of Mining Engineering developed a new technology-driven curriculum, which included more innovation and future technologies content into the program. The team identified that a future undergraduate mining engineering curriculum should develop the comprehension and ability to operate within a holistic nature of mining engineering across three dimensions:

1. Technological competences related to the changes caused by industry 4.0 on mining processes and people;
2. Comprehension about skills, techniques, and best practices related to environmental and social aspects impacting the license to operate; and
3. Soft skills related to social intelligence, emotional intelligence, and leadership that is required to work in an increasingly interdisciplinary environment effectively.

After a benchmarking exercise that showed that none of the mining engineering programs worldwide included the three dimensions, the group decided to move some courses to different semesters or merge them in order to make place for new courses/content. The new mining engineering program includes some of the following new courses: Digital Technologies and Mine Data Analytics; Engineering Services for Mining; Mine Transportation, Automation and Robotics; Water, Energy and the Environment; Mine Management Principles and Entrepreneurship; and Health, Safety and Mining Law. Furthermore, other courses will also have their content reviewed to take into account Industry 4.0 [36].

Some examples of universities that have already started to adapt their curricula are Curtin University [37], University of New South Wales, and the University of Queensland [38].

According to the Mines of the Future report, mining engineering education efforts must include research and development engineers with the following essential qualifications:

- High-level technical skills;
- Knowledge and ability to apply, optimize, and adapt to emerging technologies, especially digital technologies;
- Data literacy and capacity of manipulating large datasets (sometimes Big Data) to manage efficiently and control systems;
- The capacity of planning and operating mines with more socially acceptable surface footprints and environmental issues;
- Holistic knowledge about the full value chain of the mining operation, incorporating systems-approach to planning and operations;

- Use of risk-based techniques for planning, decision-making, and management; and
- The capacity of working as a team member or leader of multidisciplinary groups.

Although some of the current academic programs already cover a portion of these skills, it is essential to encourage universities to embrace significant changes to ensure that they are efficiently preparing mining professionals for the future needs of the industry [8].

A continuous learning cycle will characterize the mine of the future. Mining companies will become a learning organization that recognizes and values lifelong learning. Learning and personal development will be integrated into the ordinary mining engineering work. This will demand a different approach for organizational structure, with a high degree of decentralization of both authority and responsibility [39].

In-House Training and Qualification

It is important to remember that future employees are not necessarily universities or high school students with the qualifications required for these roles. The mining recruitment system will not be able to hire all the new skills required. The quickest and most cost-efficient solution relies on looking internally and developing their current talent. Many have acquired new skills, finished training, and adapted their skills to play a new role. A Pew Research Centre survey, published in 2016, found that 87% of American employees consider it will be necessary for them to get training and develop new job abilities during their employment life to stay up to date with modifications in the workplace [40]. Some major companies are already pushing this trend. For example, Amazon recently made a $700 million commitment to reskilling 100,000 workers to enhance their technology-based skills (e.g., warehouse workers were trained to become basic data analysts). JPMorgan Chase decided to invest $350 million over a five-year period to develop technical expertise in high-demand skills. One of the most important investments has been made by Walmart, which spent more than $2 billion in wages and training programs. In the Walmart Pathways program, entry-level employees learn corporate business model principles and essential soft skills [41].

Mining companies also need to know how to continue to learn and retrain their staff to confidently use modern and dynamic technologies like AA and automated systems. Only a limited number of external experts with the appropriate knowledge will be available , and businesses will actively develop new skills to avoid shortages. Relevant skill-based training should consider technical and soft competencies to fulfill the demand for collaborative, innovative, and system-thinker individuals. They can deal with high-complexity issues and see the interconnectedness and improvement across the value chain [25]. Technology can be used to boost training in work, like Virtual Reality or improved mobile devices. Mobile and wearable devices direct workers in new jobs, both on-site and remotely.

The competition to be employers of choice in the mining industry is growing. Provided the competitive market for data-related skills, it might be hard to get into the mining industry considering the negative public perception, hard work conditions, and general aspects associated with the sector's brand. So, how to attract young talent to choose to work in a remote area when they can work in the Silicon Valley or a modern office in a large and developed city?

In order to win this battle, organizations need to clarify business goals and align them with their talent strategy. This includes identifying specific tasks that could be disrupted by technology in each role now or in the future [20]. This also involves redesigning their talent acquisition strategies, changing the reliance on bonuses and travel perks approach, and becoming more open to competing for workers that prioritize flexibility in the urban life [19]. Instead of relying on a fixed schedule, the work dynamic is moving to a more flexible model (anywhere and anytime). The focus now is on a more agile and experimental approach, adaptative learning, outcomes, and fostering oriented cultures. It demands an environment where people can innovate from the bottom, changing the focus from resource-orientation to people, human capital, and brainpower. Management must be prepared to upskill employees to work on more complex tasks and providing learning, access to onboarding, and development opportunities to attract and retain top performers [19]. The retention strategy can be expanded to include the physical, emotional, financial, and spiritual health and well-being of employees [19].

Location of Future Work

The impacts of digitalization and technology advances will not only be noticed in the professional profile and skills required in the mine of the future. The workflows and work location will also change dramatically. There is a more dynamic and decentralized infrastructure and process for this working space, with new models for cooperation and interaction with customized technology, which will allow more mobility and accessibility [8].

Remote Operation Centers

The off-site environment provided by the Remote Operation Centers centralizes and connects operations in the mines and plants, protecting employees from the risks of the on-site operation and reducing costs. The remote work may also be an alternative to attract more talents to work on mining organizations. Such control rooms can be found almost everywhere in the world due to improvements in communications systems, advanced sensing and systems procedures, navigation, and imaging technologies [24]. There will be less need for operators to manage, assess, and improve operations, but more skilled professionals [8]. Technology-aided workers may serve two or three traditional roles with a focus on control, problem-solving, and organizational improvement. A single worker can track a certain process in several mine sites (e.g., truck dispatch) [42]. Computer programs with algorithms can more efficiently control fleets of self-contained carriages and loads. Driverless technology can increase output by 15%–20%, fuel consumption by 10%–15%, and maintenance costs by an 8% decrease [27].

The remote operation centers also facilitate collaboration across organizational boundaries by making it easy to connect with other internal processes and even remote parties. This integration enhances the potential for improved efficiency where discrete parts of the value chain, such as a drill, blast, loading, transporting, and processing, can be adjusted to meet business objectives.

On-Demand Experts

The mine of the future will also be marked by a more intense collaboration and integration with internal and external partners. With the gap in specialized skills, there will be a trend for services provided by on-demand experts. Indeed, some areas such as the long-term mine planning have already been applied to this type of specialized external support for many years, since there is a lack of specialists in this field and the frequency of activities does not justify a fixed professional in operation. The current disruptive scenario opens an avenue for mining and data analytics consulting firms located in large cities, empowered by a range of high-end systems to carry out various parts of the job. These companies take advantage of having many experts working with peers and sharing experiences from multiple clients.

Additionally, this new landscape is strengthened by emerging technologies, such as virtual networks and real-time connectivity with the workforce in the mine sites. Situations that require rapid changes, e.g., market opportunity, can be quickly addressed by connecting experts in different locations and using appropriate technology to facilitate their interaction [42]. The changes are not restricted to low-demand activities. For example, in the future of field operations, maintenance technicians will be armed with augmented reality tools and connected with teams of experts at headquarters that will guide them through the complex repairs in a step-by-step manner. Even better, other critical conditions can be identified and proactively fixed by the team. This work dynamic enables greater awareness of the status of each part of an operation, enhancing the integration of information available and providing a view of the whole operation [42,43]. Consequently, a much faster and efficient resolution is obtained.

Summary

The application of AA solutions has been transforming organizations in many industries. A data-driven company should be prepared for extracting the maximum value from data by focusing on three strategic challenges: corporative culture, talent acquisition and retention, and technology. Significant factors to enhance analytical maturity and accelerate companies' AA journey involve leadership engagement, well-designed data governance structure, system integration, and implementation of new technologies for extracting, storage, and processing of data across the mining value chain. Similar to many other industries, technology advances will dramatically change many occupations in the sector, and mining companies have to adapt their workforces' skill sets in order to keep pace in this very fast-changing world. Data literacy, understanding about digital technologies and systems, and soft skills such as emotional intelligence are valuable qualifications to succeed in this modern mining scenario. Some strategies to be adopted by mining companies may include a partnership with research centers, mining schools and universities, in-house training, and incorporation of outsourcing specialized services.

References

1. Diaz, A., K. Rowshankish, and T. Saleh, Why Data Culture Matters. 2018 [cited 2019 28/12/2019]; available from: https://www.mckinsey.com/business-functions/mckinsey-analytics/our-insights/why-data-culture-matters.
2. Bean, R. and T. Davenport, Companies Are Failing in Their Efforts to Become Data-Driven. 2019 [cited 2019 28/12/2019]; available from: https://hbr.org/2019/02/companies-are-failing-in-their-efforts-to-become-data-driven.
3. BHP. Inside the BHP Innovation Centre. 2019 [cited 2019 28/12/2019]; available from: https://www.bhp.com/our-businesses/technology/inside-the-bhp-innovation-centre.
4. Vale. Vale Artificial Intelligence Center. 2019 [cited 2019 28/12/2019]; available from: http://www.vale.com/en/aboutvale/news/pages/vale-inaugurates-artificial-intelligence-centre.aspx.
5. Tinto, R., Using data to find solutions, one byte at a time. 2017 [cited 2019 28/12/2019]; available from: https://www.riotinto.com/ourcommitment/spotlight-18130_24562.aspx.
6. Gold, B., Using Analytics to Enhance Decision-Making. 2017 [cited 2020 15/01/2020]; available from: https://www.barrick.com/news/news-details/2017/Using-Analytics-To-Enhance-Decision-Making/default.aspx.
7. Anglo American, Using Data Analytics to Build a Safer Workplace: The OIS. 2018 [cited 2019 15/12/2019]; available from: https://www.angloamerican.com/

futuresmart/our-world/safety-and-our-people/using-data-analytics-to-build-a-safer-workplace-the-ois.

8. Niu, R., Zulch, P., Distasio, M., Chen, G., Shen, D., Wang, Z., & Lu, J. (2019, March). Joint-Sparse Decentralized Heterogeneous Data Fusion for Target Estimation. In *2019 IEEE Aerospace Conference*, pp. 1–10. IEEE.

9. Delaware. The Data-Driven Organization: Building a Data Strategy. 2019 [cited 2020 23/01/2020]; available from: https://www.delaware.pro/en-be/solutions/data-driven-organization.

10. Davenport, T.H. and R. Bean, Big companies are embracing analytics, but most still don't have a data-driven culture. *Harvard Business Review*, 2018. 3: pp. 6–8.

11. Brown, B. and J. Gottlieb, The Need to Lead in Data Analytics. 2016 [cited 2020 23/01/2020]; available from: https://www.mckinsey.com/business-functions/mckinsey-digital/our-insights/the-need-to-lead-in-data-and-analytics.

12. Picchi, A., The Best Jobs in America for 2019. 2019 [cited 2020 22/01/2020]; available from: https://www.cbsnews.com/news/the-50-best-jobs-in-america-for-2019-according-to-glassdoor/.

13. Gottlieb, J. and A. Weinberg, Catch Them If You Can: How Leaders in Data and Analytics Have Pulled Ahead. 2019 [cited 2020 14/01/2020]; available from: https://www.mckinsey.com/business-functions/mckinsey-analytics/our-insights/catch-them-if-you-can-how-leaders-in-data-and-analytics-have-pulled-ahead.

14. Barton, D. and D. Court, Making Advanced Analytics Work for You. 2012 [cited 2020 10/01/2020]; available from: https://hbr.org/2012/10/making-advanced-analytics-work-for-you.

15. Deloitte. Tracking the Trends 2018 - The Top 10 Issues Shaping - Mining in the Year Ahead. 2018 [cited 2020 10/01/2020]; available from: https://www2.deloitte.com/content/dam/Deloitte/us/Documents/energy-resources/us-er-ttt-report-2018.pdf.

16. McKinsey & Company, Behind the Mining Productivity Upswing: Technology-Enabled Transformation. 2018 [cited 2020 23/01/2020]; available from: https://www.mckinsey.com/industries/metals-and-mining/our-insights/behind-the-mining-productivity-upswing-technology-enabled-transformation.

17. PWC, We Need to Talk About the Future of Mining. 2017 [cited 2020 23/01/2020]; available from: https://www.pwc.com.au/publications/we-need-to-talk-about-future-of-mining-2017.html.

18. PWC, Workforce of the Future - The Competing Forces Shaping 2030. 2018 [cited 2020 22/01/2020]; available from: https://www.pwc.com/gx/en/services/people-organisation/publications/workforce-of-the-future.html.

19. Deloitte. Future of Work in Mining: Attracting, Developing and Retaining Talent. 2018 [cited 2020 22/01/2020]; available from: http://www3.weforum.org/docs/WEF_Future_of_Jobs_2018.pdf.

20. Deloitte. The Top 10 Issues Shaping Mining in the Year Ahead. 2019 [cited 2020 10/01/2020]; available from: https://www2.deloitte.com/content/dam/Deloitte/tr/Documents/technology-media-telecommunications/Fast50-2019-Turkey-Report.pdf.

21. Lööw, J., L. Abrahamsson, and J. Johansson, Mining 4.0—The impact of new technology from a work place perspective. *Mining, Metallurgy & Exploration*, 2019. 36(4): pp. 701–707.

22. Goidel, G., Data Is the New Oil: Rio Tinto Builds New Intelligent Mine. 2018 [cited 2020 24/01/2020]; available from: https://digital.hbs.edu/platform-rctom/submission/data-is-the-new-oil-rio-tinto-builds-new-intelligent-mine/.
23. CSIRO. Downhole Data without Delay. 2018 [cited 2020 24/01/2020]; available from: https://www.csiro.au/en/Research/MRF/Areas/Resourceful-magazine/Issue-10/Downhole-data-without-delay.
24. Gleeson, D., Vale after More Cost Savings with Development of Artificial Intelligence Centre. 2019 [cited 2020 12/01/2020]; available from: https://im-mining.com/2019/01/10/vale-cost-savings-development-ai-centre/.
25. Ernst and Young, Future of work: The economic implications of technology and digital mining. 2019.
26. Romero, D., et al., Towards an operator 4.0 typology: A human-centric perspective on the fourth industrial revolution technologies. In *Proceedings of the International Conference on Computers and Industrial Engineering (CIE46)*. 2016. Tianjin, China..
27. Cosbey, A., Mann, H., Maennling, N., Toledano, P., Geipel, J., & Brauch, M. D. (2016). Mining a mirage: Reassessing the shared-value paradigm in light of the technological advances in the mining sector.
28. Horberry, T., R. Burgess-Limerick, and L.J. Steiner, *Human Factors for the Design, Operation, and Maintenance of Mining Equipment*. 2016: CRC Press, Boca Raton, FL.
29. Minerals council of Australia, MTEC Key Performance Measures Report 2018. 2018 [cited 2020 24/01/2020]; available from: https://minerals.org.au/news/mtec-key-performance-measures-report-2018.
30. Knights, P.F., Short-Term Supply and Demand of Graduate Mining Engineers in Australia. 2019 [cited 2020 24/01/2020]; available from: https://link.springer.com/article/10.1007%2Fs13563-019-00208-0.
31. Kazanin, O.I. and C. Drebenstedt, Mining education in the 21st century: Global challenges and prospects. Записки Горного института, 2017. 225.
32. Scoble, M. and D. Laurence, Future mining engineers: Educational development strategy. In *Proceedings of the International Conference on Future Mining*. 2008: Citeseer, Sydney.
33. Diana, D., Global mineral engineering curriculum review. *Mineral Processing Education Around with World*, 2018. 1(3): pp. 61–75.
34. SAP, The future of mining for the betterment of industry and environment. SAP Industries White Paper | Mining, 2019. 1(1).
35. Jones, O., Digital Mines: The Need for Restructuring University Courses. 2018 [cited 2020 25/01/2020]; available from: https://www.ausimmbulletin.com/feature/digital-mines-need-restructuring-university-courses/.
36. Mitra, R., et al., Curriculum review process at the school of mining engineering at the University of the Witwatersrand. *International Journal of Georesources and Environment-IJGE (formerly Int'l J of Geohazards and Environment)*, 2018. 4(3): pp. 54–58.
37. Philips, Y., New Curriculum to Train Resources Workforce of the Future. 2019 [cited 2020 22/01/2020]; available from: https://news.curtin.edu.au/mediareleases/new-curriculum-to-train-resourcesworkforce-of-the-future/.
38. X'Abbey, E., Digital Mines: On the Pathway to Mining without Miners. 2017 [cited 2020 24/01/2020]; available from: www.spatialsource.com.au/company-industry/digital-mines-pathway-mining-without-miners.

39. Johansson, B., J. Johansson, and L. Abrahamsson, Attractive workplaces in the mine of the future: 26 statements. *International Journal of Mining and Mineral Engineering*, 2010. 2(3): pp. 239–252.
40. Pew Research Center, The State of American Jobs. 2016: Pew Research Center, Washington, DC.
41. Hancock, B., K. Lazaroff-Puck, & S. Rutherford. Getting practical about the future of work. *The McKinsey Quarterly*, 2020. (2): pp: 123–132.
42. Bassan, J., V. Srinivasan, P. Knights, and C. Farrelly, A Day in the Life of a Mine Worker 2025. 2008 [cited 2020 22/01/2020]; available from: https://espace.library. uq.edu.au/view/UQ:176886.
43. Benjamin, G., B. May, M. Prema, and V. Raghudanshi, The Coming Evolution of Field Operations. 2019 [cited 2020 24/01/2020]; available from: https://www.mckinsey.com/business-functions/operations/our-insights/the-coming-evolution-of-field-operations.

Index

Note: **Bold** page numbers refer to tables and *italic* page numbers refer to figures.